Understanding and Using Programmable Controllers

Understanding and Using Programmable Controllers

Thomas E. Kissell

Terra Technical College

Prentice-Hall, Inc., Englewood Cliffs, New Jersey 07632

Library of Congress Cataloging-in-Publication Data

KISSELL, THOMAS E.
    Understanding and using programmable controllers.

    Includes index.
    1. Programmable controllers.   I. Title.
TJ223.P76K57 1986   629.8'95   85-19095
ISBN 0-13-937129-X

Editorial/production supervision
and interior design: **Kathryn Pavelec**
Cover design: **Whitman Studio, Inc.**
Manufacturing buyer: **Rhett Conklin**

Printed in the United States of America

10   9   8   7   6   5   4   3   2

ISBN 0-13-937129-X   01

PRENTICE-HALL INTERNATIONAL (UK) LIMITED, *London*
PRENTICE-HALL OF AUSTRALIA PTY. LIMITED, *Sydney*
PRENTICE-HALL CANADA INC., *Toronto*
PRENTICE-HALL HISPANOAMERICANA, S.A., *Mexico*
PRENTICE-HALL OF INDIA PRIVATE LIMITED, *New Delhi*
PRENTICE-HALL OF JAPAN, INC., *Tokyo*
PRENTICE-HALL OF SOUTHEAST ASIA PTE. LTD., *Singapore*
EDITORA PRENTICE-HALL DO BRASIL, LTDA., *Rio de Janeiro*
WHITEHALL BOOKS LIMITED, *Wellington, New Zealand*

To my wife Kathleen
and my children Kelly and Christopher

Contents

Preface

This text is a comprehensive view of programmable controllers. It starts at an introductory level with the assumption that the reader has no previous experience on P.C.'s. Any knowledge the reader has about motor controls and digital electronics would be helpful, but is not necessary.

The author has carefully chosen programming examples that will reflect formats of the most widely used P.C.'s, since a format for every P.C. in use today cannot be presented in one book. It is the author's intention that the programming methods presented will provide the reader with the information needed to program similar types of P.C.'s.

Before reading this text, it should be noted that it was not written to replace the technical data that is supplied by P.C. manufacturers. Rather, this text should give the reader an essential understanding of basic P.C. theory and operation so that the manufactured material can be fully utilized.

The first four chapters of this text cover basic information concerning all P.C.'s. This information includes basic P.C. operation, history, numbering systems, and programming panels.

The next five chapters include programming examples to cover typical P.C. operations such as coils and contacts, timers, counters, sequencers, and math functions. These chapters explain how the major P.C. brands are formatted and programmed.

The next seven chapters give the reader an understanding of installation, troubleshooting, maintenance, and the operation of hardware like I/O modules and power supplies.

The last four chapters discuss advanced uses of P.C.'s for such things as acquisition and process control. Also included in this section are chapters on program documentation and advanced I/O modules.

Each chapter includes diagrams and pictures to help the reader understand P.C.'s. All chapters have questions at the end, and in some cases, programming exercises as well. The author hopes that the reader will try some or all of the programming examples to enhance his or her understanding of programmable controllers.

The author wishes to acknowledge the following people who supplied pictures and product information: Mr. Howard Hendricks from Gould Inc. Programmable Controller Division; Mr. Christopher Roy from Allen-Bradley; Ms. Phyllis Barr Fox from Texas Instruments; and Mr. Michael Waletzko from Square D Company.

I would also like to thank my wife Kathleen for all the work she put into typing this manuscript; and my children Christopher and Kelly, for their help in assembling parts of this material.

Thomas E. Kissell

Understanding and Using Programmable Controllers

chapter 1
Introduction

Programmable controllers, or P.C.'s, have been used in industry in one form or another for the past fifteen years. The newest models may not resemble their earlier counterparts, but many of the concepts used in the first units are still in use. The picture in Fig. 1-1 shows a typical programmable controller. The programmable controller is basically a computer-controlled system containing a microprocessor that is programmed with a programming panel, or keyboard. The P.C. is dedicated to receiving input signals and sending output signals in response to the program logic. The program generally consists of contacts, outputs, timers, counters, and math functions. The programmable controllers found in industry today have evolved from the need for a control system that can be easily reprogrammed as changes occur or as new products develop.

For example, the automotive industry is faced annually with major changes in production as new models are designed. This changeover previously required electricians and maintenance personnel to put in long hours to rewire relay-type controls. Each changeover period was costly to the industry, and it often forced changes to be infrequent and as simple as possible.

The programmable controller was developed in 1969 to ease the problem of changing control systems periodically. Modern P.C.'s consist of four major parts: a central processing unit, also known as the processor or computer; an input section; an output section; and a power supply (see Fig. 1-2). The programming panel is not considered a major part of an operating P.C. because once you have used it to program the processor, it can be disconnected and the system will still operate correctly. The relationship of these four parts and the ability of the P.C.'s to be

1

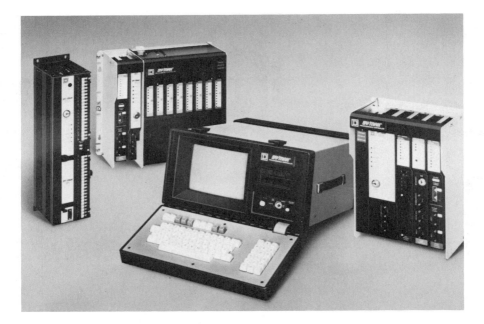

Figure 1-1 A typical P.C. system. Picture courtesy of Square D Company.

reprogrammed can best be understood by comparing hardwired circuits to identical circuits controlled by a P.C.

From Fig. 1-3 you can see that Switch 1 and Switch 2 are normally opened push-button switches. Switch 1 will send power to Lamp 1, and Switch 2 will send power to Lamp 2. When Switch 1 is pushed closed, Lamp 1 will light. When Switch 2 is closed, Lamp 2 will light.

Figure 1-4 shows the same components connected to a P.C. From this diagram you can see several differences. First, switches are not connected directly to the lamps, instead, the *switches* are connected to *input modules*, and the *lamps* are connected to *output modules*. Another difference is that the input modules and

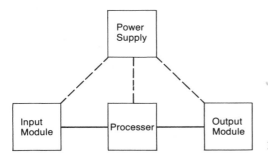

Figure 1-2 Block diagram of a P.C.

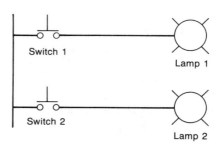

Figure 1-3 Hardwired system.

output modules are not connected to each other directly. They are connected through the *processor* when the program logic indicates certain conditions have been met.

The processor in Fig. 1-4 is programmed to connect Switch 1 to Lamp 1, and Switch 2 to Lamp 2. This is accomplished by typing a program or diagram into the processor from a keyboard. The program looks very similar to a regular electrical ladder diagram.

The operation of the hardwired switches and lamps and the P.C.-controlled system seem identical. When Switch 1 closes, Lamp 1 lights, and when Switch 2 closes, Lamp 2 lights. The major difference is in the way electricity flows to accomplish this.

In the hardwired system, the electrons flow from the voltage source, through the switch, to the correct indicator lamp. Electrical power simply follows the wire conductors to the lamp. When the switch is opened, power is interrupted and the light goes out.

In the P.C.-controlled system, electrical power comes from the voltage source, through the switch, into the input module. The input module senses the presence of this voltage and in turn, sends a small signal voltage into the processor. The voltage from the switch is isolated from the voltage signal that the module sends

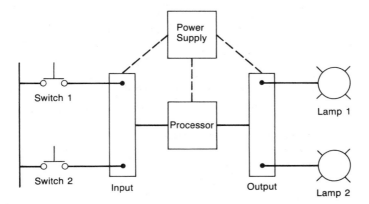

Figure 1-4 P.C. system.

into the processor. This isolation is necessary since the fragile processor chip operates at very low voltage and current levels. Isolation is generally provided by an electronic component known as an *opto coupler*, which enables the power supply to provide the appropriate voltage levels to input modules, output modules, and the processor.

The processor receives a signal from the input module when the switch is closed, and will send a similar signal to the output module as directed by the program. The program directs the processor-transmitted signal to the appropriate output terminal for Lamp 1 when it receives a signal from the input module terminal connected to Switch 1. All of this takes place in less than a thousandth of a second.

When Switch 2 is activated a similar action is completed by the processor, but this time the signal is sent to the output module terminal for Lamp 2.

An observer of both hardwired and P.C.-controlled systems would not notice any difference in system operation. In both systems, Switch 1 controls Lamp 1, and Switch 2 controls Lamp 2.

The biggest advantage of the programmable controller becomes evident when a change is needed in the circuits previously discussed. For example, if you needed to change the circuits of a hardwired system to have Switch 1 control Lamp 2, and Switch 2 control Lamp 1, it would take nearly ten minutes to rewire them, and would involve exchanging the wires at the switches or the lamps.

The same change in a P.C.-controlled system is accomplished by reprogramming the P.C. Since changes can be made in the program instead of the wiring system, many man-hours are saved.

This simple example of changing a program can be applied to a system containing fifty or more switches and outputs. You can see that by merely editing or changing the program, a great deal of time can be saved.

Now that you can see the major advantage of using a programmable controller, we can proceed with more information about the basic system. In the previous example, the P.C. used a program to determine which lamp to turn on. This program is entered into the processor through the keyboard, stored in the processor's memory, and displayed on a screen called a *CRT* (cathode-ray tube). The CRT is similar to the tube used in televisions and allows the P.C. programmer to view the present diagram that is in the processor's memory. The program on the CRT screen looks very similar to a regular electrical diagram and uses normal electrical diagram symbols. Another feature of most P.C.'s is their capability of being connected to a printer, so that a hard-copy printout can be made of the electrical diagram. The hard-copy printout of the program is used for troubleshooting.

Other usable features on most programmable controllers include the availability of timers and counters with programmable times and counts. These timers and counters can be programmed to simulate the operation of electromechanical timers and counters. The maximum number of counters and timers in a program depends on the size of processor's memory. The processor chip also gives P.C.'s the normal features of other computers, such as addition, subtraction, multiplication, and division functions.

MORE ABOUT PROGRAMMABLE CONTROLLERS' BASIC PARTS

Figure 1-4 shows the input and output sections of the programmable controller as rectangular boxes. The boxes are used to show simplicity. Figure 1-5 shows a better view of the input and output sections of a P.C. You can see that the input and output modules are connected side by side in the I/O module housing. At first, all you need to know about input and output modules is that they are needed to receive signals from switches and send signals to output coils and lamps, as directed by the processor. In later chapters you will become more familiar with their internal operation and how the processor identifies them.

You will also find that even though most P.C. input, output, and specialty modules all look alike, they have a variety of functions. They are also wired in a variety of ways. Some differences will even be apparent between various models made by the same manufacturer.

The input and output section is used to interface the P.C. with many industrial operations and machines. This means each input and output listed in the program must be given a number for identification. This number is also used as a storage location for the program device inside the processor memory and as an address location for the input or output module in the housing. At first, this may seem a little confusing, but you will soon understand that every component in the program must be numbered so you can identify it. The processor uses the address to send and receive signals to the modules. The processor must also use the identification number to remember where data pertaining to the programmed device is stored in memory. Finally, you will use the identification number as the address to physically

Figure 1-5 Picture of I/O modules and processor. Picture courtesy of Gould Modicon Programmable Controller Division.

locate the input and output modules in order to wire and troubleshoot them. Chapter 5 discusses this numbering system in depth. Until then you will only need to be aware that the numbering system exists and that it is vital for the P.C. system to operate correctly. You will also learn that each manufacturer designs its own numbering system. All these systems are somewhat similar and easy to learn.

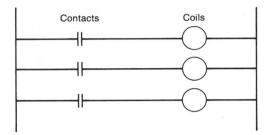

Figure 1-6 Ladder diagram format.

AN EXAMPLE PROGRAM
FOR PROGRAMMABLE CONTROLLERS

The basic program format for programmable controllers is very similar to an electrical ladder diagram. The ladder diagram is so-called because the lines of a completed diagram resemble the rung of a wooden ladder (see Fig. 1-6). In fact, some P.C. manufacturers like Allen-Bradley refer to each program line as a rung.

The ladder diagram is constructed to show the sequence of events, rather than the path of wires and the layout of the electrical parts. Figure 1-7 shows a typical start-stop circuit controlling a motor starter. The circuit is drawn as a wiring diagram that shows component placement and wiring paths in Fig. 1-7 and as a ladder diagram that shows the sequence of operation in Fig. 1-8.

You should notice that the wires in the wiring diagram may cross one another and become hard to follow, even though the circuit has only three components. This happens because accurate location of components is the main function of the

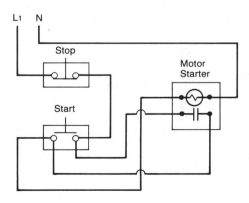

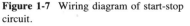

Figure 1-7 Wiring diagram of start-stop circuit.

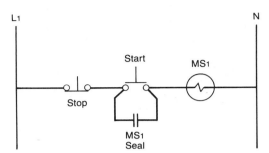

Figure 1-8 Ladder diagram of start-stop circuit.

wiring diagram. You can imagine how many wires might cross if the diagram was showing fifty components.

In comparison, you should notice that in the ladder diagram the wires should never cross. Power is provided to each rung of the diagram from the vertical line that runs down the left side of the diagram. A vertical line also runs down the right side of the diagram to provide power to complete the circuit. These two vertical lines are usually called L1 and L2 for line 1 and line 2, or L1 and N for line 1 and neutral. These are the terminals where power is provided from the transformer. Each line, or rung, generally has one coil, or load, and can have as many switches or contacts as is required to control the load. Some P.C.'s will limit the total number of contacts and coils in each program line or rung.

Another important difference that you should notice when looking at the wiring diagram in Fig. 1-7 and the ladder diagram in Fig. 1-8 is that the coil of the relay, or motor starter, is shown very near the contacts it is controlling. This is because the coil needs to be physically very close to its contacts so that its magnetic force can open or close its contacts. Remember the wiring diagram is drawn to show the true location of each part, or component. In comparison, the coil in a ladder diagram may show up several rungs or even pages away from the contacts it controls because it sequentially occurs this way. In other words, the contacts of a relay or motor starter may show up anywhere in the ladder diagram. This may seem difficult to follow at first, but each contact is identified by a number or name to tell which coil operates it. All you need to remember is that the coil must be energized before its magnetic field can move the contacts. It may be easier if you make a note right above or near each contact set you find in a program, indicating the name of the relay coil that controls it. That way you won't become confused by finding contacts of one relay in several different program lines.

The typical programmable controller program looks very similar to a ladder diagram. This format is used because the ladder diagram has been the working language of electricians for many years. This format style is also used because the computer scans the program in a sequential manner. Some early programmable controllers tried to use BASIC, Pascal, and FORTRAN computer languages to control inputs and outputs. Even though these were commonly used computer languages, they did not succeed as languages for programmable controllers because electricians could not understand them to troubleshoot and repair large industrial

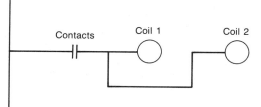

Figure 1-9 Two coils in series.

systems. Since programmable controllers now use the ladder diagram-type program, they have truly become an electrician's control system.

The actual programming formats for each programmable controller may be a little different from each other, but they all resemble the original electrical ladder diagram. Most systems still refer to the vertical line running down the left side of the program as the source for all power to each program line or rung. Some manufacturers even refer to this line as the power rail. For this reason, when you connect contacts and coils in a circuit, the beginning of that circuit must touch the left side of the program to get power. The format for some P.C.'s requires you to start each program line at the power rail and then work to the right side. They also require that you put at least one load in each circuit. This load will usually be a coil, but you could use timers, counters, or other similar function blocks for the load. Some brands of P.C.'s allow you to put several loads on the same line. This appears to put the loads in series. All electricians know that loads, such as coils, will not function correctly if they are connected in series. What has really happened in the P.C. is that solid state electronic chips accommodate this in the processor's program, but the coils or outputs actually operate as though they are in parallel. In other words, if two coils are in series in the program as shown in Fig. 1-9, they operate as though they are in parallel.

If the first coil does not operate for some reason, the second one will still function. This is one major difference between the electronic outputs in a programmable controller, and electromechanical devices, such as coils and solenoids, in a hardwired system. As you complete this text, you will notice several other differences like this that allow you to do time-saving functions on the programmable controller. Remember, some P.C.'s will not allow two coils or loads on the same program line, so be sure to check the format of your P.C.

Another programming feature that tends to bother someone trying to learn the programmable controller for the first time is the method that P.C. manufacturers use to explain how their program is scanned in memory. This may be accomplished by the processor starting at the top left side and scanning each column from top to bottom, then moving to the next column and scanning it from top to bottom. Other processors may scan rung by rung. Remember the computer chip, or processor, has its own method for entering, storing, and executing a program. You will only need to know the general operation of the processor, since some specific instructions will not be functional due to the type of scan.

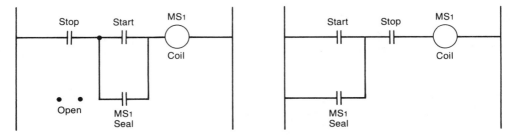

Figure 1-10 Use of horizontal open or moving seal contacts.

For example, the way the processor scans the program on your P.C. definitely determines where you can leave spaces in the program and how those spaces are treated. In Fig. 1-10, you see the basic start-stop circuit with the seal contacts around the start contacts. Some processors require that you either put in a space symbol called a *horizontal open*, or relocate the start contacts so the seal contacts in the second line are placed right against the power rail. Both methods are illustrated in these diagrams. If you put the seal contacts against the power rail, you must also reposition the start contacts against the power rail in the top line so that the seal contacts remain in parallel with them. The two circuits in Fig. 1-10 operate exactly the same, you merely have to accommodate the processor's method of scanning the program. This is why all P.C.'s are not programmed exactly alike.

Another important point that you have probably noticed by now is that the stop switches in the circuits shown in Fig. 1-10, and other circuits, have been programmed as normally open contacts. You probably feel this has been a mistake or a misprint, but in reality, the stop switch is programmed as normally open in *all* P.C. programs.

As you know, the stop switch is actually a normally closed push-button switch. You may better understand why the normally closed stop switch shows up in the program as an open switch from the following example. Figure 1-11 shows a diagram

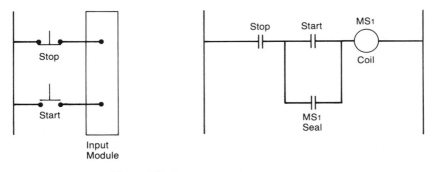

Figure 1-11 Start-stop circuit and I/O modules.

Figure 1-12 A large-type programmable controller. Picture courtesy of Gould
Modicon Programmable Controller Division.

of the stop switch connected to an input module, and the program of a start-stop
circuit.

From the program, you can see that the output coil will be energized when
both the start and stop switches pass power. In order for the open contacts (start
switch) in the program to close, an input signal must be received at the input module
from the start switch. The input signal will be received from the start switch when
someone pushes the start switch closed. The signal from the stop switch will be
received by the module at all times because a spring holds the stop switch contacts
closed. In fact, the input signal for the stop switch will be present at the module
until someone pushes on the stop switch and opens these contacts.

You can now see that if the stop switch were programmed as a closed switch,
the circuit would not operate correctly because the spring in the stop switch would
close the contacts, causing the stop contacts in the program to be opened and
preventing power from being passed to the output.

You should now begin to see that some things that seem to be contradictory
at first will have a very simple explanation. If you will keep these things in mind

as you read on in this text, you will see that some programmable controllers are programmed in a similar manner or have similar functions because they use the same processor chip. Even the larger P.C. systems, like the one shown in the picture in Fig. 1-12, are programmed similarly. Other P.C.s, even though they are made by the same company, are programmed slightly differently because a different processor chip can be used in each model. You will become more comfortable with these minor differences as you gain programming and operational experience.

QUESTIONS

1. What is a programmable controller?
2. Why was the first programmable controller made?
3. Name the three basic parts of a programmable controller.
4. Explain the difference between a wiring diagram and a ladder diagram.
5. Explain the advantages a programmable controller has over a similar hardwired control system.
6. If five sets of contacts and three coils show up in one ladder diagram, how will you know which coil will control each set of contacts?
7. Explain how the processor scans a program.
8. Why is a normally closed stop switch always programmed as a set of normally open contacts?
9. Explain why contacts and coils must be numbered in a P.C. program.
10. Explain how the processor uses the number assigned to the contacts or coils.
11. Explain how a technician would use the numbers assigned to the contacts or coils.

chapter 2
History

As you begin to use programmable controllers, you may wonder when they were first designed and used. The P.C. industry is relatively young, since the first systems have only been in use since the late 1960s and early 1970s. You will begin to get an understanding of the evolution of P.C.'s when you see the growth the major manufacturers have made in the past years.

MODICON P.C'S

The earliest fully programmable P.C. was developed in 1969 by a consulting engineering firm called Bedford Associates. Bedford Associates later changed their name to Modicon, and their first P.C. was designed as a dedicated computer control system built specifically for the General Motors Hydramatic Division. The automotive industry put out the call to anyone interested in developing this type of system. In June 1969, Bedford Associates delivered a P.C. system called the 084 to the General Motors Hydramatic Division. This first system, also called a *Hard Hat*, was named the 084 because it was the eighty-fourth system to be developed after eighty-three prior modifications. More modifications were made, and several other models were produced with more specialized characteristics through the early 1970s. These models included the 184, which was developed in 1972, and the 384, which was developed very soon thereafter. The 184 and 384 looked and operated like the P.C.'s used today. They were fully programmable, using ladder logic diagrams and utilizing solid-state design for I/O with a microcomputer processor. These early systems were fully compatible with tape loader/recorders and could utilize TTL, analog voltage, analog current, and ASCII, and computer interface modules.

During this same era, Modicon also produced two other models. These models were the 284 and the 1084. The 284 was a small system that controlled 80 inputs and 40 outputs. The 1084 was capable of controlling 5120 inputs and 5120 outputs and had 40K of core memory. These systems allowed for the control of large and small systems and led the way for the model 484, which has been an industry standard since 1976.

The 484 system allows for the control of 256 inputs and 256 outputs from one processor. The 484 also provides counters, timers, and sequencers. Other enhanced functions include conversion to and from BCD and binary input and output, table-to-register and register-to-table data manipulations, and math functions with double precision registers.

In 1977, Modicon was purchased by Gould Inc. In 1978, the Modbus data network was designed to allow the 484 and other Modicon P.C.'s to transfer data to each other. The first Modbus system was installed and became operational in November 1979. In 1980, Modicon presented a small, compact, low-cost, powerful P.C. system. This system was called the Micro 84 system and was capable of controlling 64 I/Os and provided counters, timers, sequencers, and math functions.

The Modicon 584M (medium size) and Modicon 584L (large size) processors were brought onto the market along with the Modicon 884 system between 1981 and 1982. The 984 was introduced in 1984. These new P.C.'s were designed especially for process control applications, including PID. The I/O sections provided an interface to analog systems and were totally compatible with the Modbus data network. The Modvue color graphics system was also incorporated with these processors in 1982 to provide total system documentation and management.

The major emphasis for these systems is compatibility with a wide range of modules. These modules include analog input, analog output, analog multiplex, reed relay, and TTL. The Modvue color graphics display also provides the unique ability of touch screen selection. Once a diagram and menu are displayed on the CRT, the programmer can make selections by merely touching the screen.

These P.C.'s provide a wide variety of control systems. Modicon's future plans include continued updating of their systems and equipment to provide larger, more flexible memory with emphasis on program scan. The data network will also be updated to provide maximum communication capability. Other advancements will be made in process control systems and data documentation.

ALLEN-BRADLEY P.C.'s

Modicon is not alone in the P.C. industry. In 1969, when the General Motors Hydramatic Division put out the call for a P.C.-type system, Allen-Bradley also responded with a P.C. system. This system was called PLC. Even though the PLC was not chosen for the General Motors plant, it proved to be a very reliable P.C.

Allen-Bradley had been working on a solid-state control system for some time. Their first solid-state control system, the PDQ, was designed in 1959. It was not reprogrammed as easily as modern day P.C.'s, but it filled a much-needed demand as a relay replacer. The PLC system provided similar capabilities to modern day P.C.'s. These capabilities included I/O control, timers, counters, and simple register manipulation. In 1970, Allen-Bradley designed the PMC programmable controller system, which was the industry mainstay until it was joined by the PLC-2 in 1975. The PMC and PLC-2 provided advanced control and mini-computer capabilities. These capabilities included math functions, networking data transfer, program documentation, and advanced analog control.

In 1979, the PLC 2/20 was introduced, and it provided advanced control systems with extensive fault diagnostics, analog modules, signal interface, and positioning controls. Most of these early systems are still in use today with additional P.C.'s added to them. They were complex enough that they did not need to be replaced as newer models came out; instead, they were simply added to as needed.

In 1981, the mini PLC 2/15 and midsize PLC 2/30 were introduced. The largest Allen-Bradley system, the PLC-3, also came onto the market in that year. The PLC 2/15 and PLC 2/30 provided many similar functions, their main difference being memory size and total number of I/Os controlled. Some of their functions included typical mathematics, as well as counters, timers, and sequencers. Advanced functions, such as word files, block transfer, data computation, data highway compatibility, and off-line and on-line programming were also included.

The PLC-3 allowed the P.C. system to control thousands of I/Os easily. It was the answer to "whole factory" control. This system allowed P.C.'s to take charge of complete factory automation instead of just one machine. The PLC-3 also provided the most advanced qualities of its time with its dual scanner modules. Up to this time, the typical P.C. processor was so dedicated that it slowed down when the number of I/Os increased. It also became overburdened when it was expected to send and receive messages for report generation while watching for I/O updates and program scans. The PLC-3 introduced two separate scanner modules. One module takes responsibility for normal P.C. scans of program and I/O updates. The other scanner is dedicated to oversee report generation. This way the PLC-3 system can be used for sophisticated process control without overworking. Its advanced file system allow the processor to store and manipulate data in binary, binary-coded decimal (BCD), integer, and floating point. This data can be moved from file to file and from one type of number system to another. The memory is also large enough to handle thousands of data words in each file. The PLC-3 is filling the need for a large, sophisticated P.C. that can still be managed and programmed by the electrician or technician on the factory floor.

In 1982 Allen-Bradley introduced the PLC-4. This system is a small P.C. that provides 20 inputs and 12 outputs. The total P.C. system, including power supply, processor, and I/Os is contained in a single rugged case.

Even though this system is small and dedicated, several of them can be connected, or daisy-chained, to communicate among themselves. They also can be

Figure 2-1 Allen-Bradley Mini PLC-2/05. Picture courtesy of Allen-Bradley Company.

connected to the data highway. The daisy chain feature allows output coils from one processor to control contacts in another processor. This feature means PLC-4s can be connected to build a small network for several machines or added on a medium-size system for one machine. The Modular Automative Controller (MAC) system was also introduced in 1982. This system provided an automation system at an inexpensive cost.

In 1984, the PLC-2/05 and SLC-100 were also introduced. The PLC-2/05 is shown in Fig. 2-1. These systems are primarily small or mini P.C.'s that provide a wide range of functions at a low price. In this way, Allen-Bradley has provided a P.C. for the largest and smallest user.

Future trends for Allen-Bradley P.C.'s call for more intelligent-type modules, faster processor operations, larger memory, and more computer-type functions.

Other trends seem to be heading in the direction of larger data highway and system communications and more process control functions. In fact, Allen-Bradley has a color graphics system that can be connected to voice modules. This system will provide a very complex and advanced P.C. system.

While Modicon and Allen-Bradley were developing their early systems, they were not alone. From their early start they built a sales and service network to become two of the larger P.C. manufacturers.

TEXAS INSTRUMENTS P.C.'s

Many other P.C. manufacturers were also busy keeping pace with industry needs. Among these other manufacturers, Texas Instruments was able to provide unique P.C. systems and gain a major portion of the P.C. market.

In 1973, Texas Instruments designed their first P.C., called the 5TI. The 5TI was sold from the mid-1970s through present day. It is a very simple P.C. with dedicated I/O system. It provided the user with easy-to-use programming and troubleshooting capabilities at a very inexpensive price. In fact there have been more 5TI P.C.'s sold than any other P.C. models.

The 5TI provided the user with counters, timers, and I/O functions. Even though the 5TI filled the typical P.C. need, industry was looking for a system to control complex processes. In the mid-1970s, the PM550 was introduced to handle process control functions. This system could handle analog and PID functions, as well as a supplemented color graphics systems.

In 1981, the 510 system was produced to meet the need for a larger P.C. system. This system also provided the user with data network capabilities, as well as data manipulations functions.

In 1982, the 520 and 530 P.C. were introduced. Each of these systems brought new features for more enhanced functions. During this time, the data network called the TIWAY was perfected and put into wide use. The 520 and 530 systems provided a large variety of P.C. functions, allowing Texas Instruments to stay abreast of the competition.

In late 1984, Texas Instruments produced its latest P.C. system, called the Model 560/565. This system was specifically designed for larger, faster processors with large memories. By adding one card the Model 560 is converted to the Model 565, which is capable of complex process control with PID loops. This system will provide the latest control technology and be totally compatible with existing systems and networks. These new units have allowed TI to be able to stay at the top with Allen-Bradley and Modicon in the P.C. market.

OTHER MANUFACTURERS

Many other P.C. manufacturers have stayed current with the needs of the industry. Even though Modicon, Allen-Bradley, and Texas Instruments have a slightly larger share of the market, other manufacturers have put millions of dollars into P.C. research and development to stay current with the market. Many of these companies have provided very unique control applications that make their product very useful for specific applications. Many times, companies looking to purchase P.C. equipment tend to make decisions based on how big the P.C. manufacturer is, rather than how well the P.C. system will fit their specific application and long range needs.

SQUARE D P.C.'s

An example of these companies is the Square D Company. Square D Company has manufactured all types of traditional motor controls and solid-state controls. In 1959, they produced a control product called the CLASS 8852 NORPAK Solid-State Control Logic. This control system allowed the user to build a control program by grouping NORPAK modules to form logic gates. These systems primarily used transistors to form the logic circuits. This system became the forerunner of their next system, which was produced in the early 1970s.

This new system, the Class 8871 Program Sequencer Control (PSC) utilized diode matrix memory, which gave the system larger memory capability and some simple math functions.

Their first true P.C. was produced in 1973. It was termed the Class 8881 Programmable Controller. This P.C. was rather typical of other P.C.'s of that age because it was reprogrammable, used a keyboard, and had a CRT. It also had capabilities of 4K, 8K, 16K, and 32K memory and large scale I/O functions.

In 1975 the Class 8873 Ladder Diagram Processor (LDP) was introduced. It used PROM memory plus diode matrix memory. It was compatible with a data logger for report generation, black and white graphics, color graphics, PID loops, and a host computer.

In 1977, the Class 8884 Sy/Max-20 programmable controller was introduced with state of the art EAROM and RAM memory. This system was capable of PID process control, color graphics, host computer interface, and distributed control. The latest Square D controller was introduced in 1982. It was called the CLASS 8020 SY/MAX. The model 100 and model 300 programming panels were compatible with it, as well as with the SY/NET data network. The Class 8020 SY/MAX also featured an ASCII keyboard and was capable of RS232C serial communications. This system also featured several intelligent modules and brought Square D P.C.'s into the modern era.

The histories of many other P.C. companies actually follow the ones previously mentioned. The reason each of these companies' P.C.'s evolved similarly is because of the demands by industry and the evolution of the processor chip. When the first P.C.'s came onto the market in the early 1970s most microprocessor chips were limited to input/output management and simple register manipulation, such as counters, timers, and storage. These early processors were basically capable of working with 8-bit data words, and they were still fairly expensive. In the mid-1970s, the 16-bit processors were coming of age, and their capabilities and functions became more like those of traditional computers. These processors had complete complex data manipulation, data files management, increased memory size, and faster scan times. In the late 1970s and early 1980, the price of microprocessors dropped dramatically. This allowed the possibility of using several processor chips in tandem as coprocessors or as master/slave relationships. It was also possible to put processor chips in almost all other peripheral devices, such as printers, recorders, and programming panels. Even modules begin to show up with on-board processors.

Each of these technical advancements brought about a new generation of P.C.'s. At the same time, as the processors were being improved, engineers were able to solve troubling problems with I/O components and isolation components found in modules. Many companies and industrial P.C. users spent agonizing hours working on the problem of module failure in the early P.C.'s. Several breakthroughs in solid-state devices, like the SCR, Triac, and transistor now provide reliable modules. In fact, many technicians agree that modules are providing less trouble than their electromechanical counterpart, the relay. In many ways today's P.C. has become a reliable partner in industrial control. It is important to realize the heritage that today's P.C.'s share with the new high-tech electronic revolution.

QUESTIONS

1. When was the first P.C. manufactured?
2. Who purchased the first P.C.?
3. What was the first model of P.C. that Modicon made?
4. List four Modicon models of P.C.'s.
5. When did Allen-Bradley manufacture their first P.C.?
6. Name four Allen-Bradley models of P.C.'s.
7. What were the features the Allen-Bradley PLC-3 provided?

8. What was the first P.C. that Texas Instruments sold?
9. What is the latest Texas Instruments P.C.?
10. What controls did Square D sell prior to their first P.C.?
11. What year did Square D produce their first P.C.?
12. Name one reason intelligent modules have become so available.

chapter 3
Numbering Systems

People usually have one of two distinct feelings about numbers and numbering systems. They either love them, or they hate them. This is usually determined by their understanding of math and numbering systems. If they understand math, then they can use and enjoy the numbering systems. If they do not understand math and numbers, they usually feel intimidated by them and avoid their use at all costs.

You must have a good understanding of decimal, binary, octal, and binary-coded decimal numbering systems, since the processors in most programmable controllers use each of these numbering systems to transmit information between their input and output modules, memory, and the processors. For this reason, the discussion about numbering systems will be at a basic and informative level, rather than through a flashy mathematical approach.

DECIMAL SYSTEM

The easiest numbering system to understand is the system you learned in grade school. It is called the decimal system, or base ten system, because it uses ten values or digits. *Deci* is also the Greek word for ten. These ten numbers start with zero having the smallest value and 9 having the highest value. The values for the decimal system are 0, 1, 2, 3, 4, 5, 6, 7, 8, and 9.

The decimal system also uses these values in various positions for weighting of the numbers. This may seem confusing at first, but stop and think of the values that are larger than nine. Since nine is the largest value in the decimal system, a column to the left of the first number must be added to count values higher than nine. For example, to get ten, you would put a 1 to the left of a 0. Now you have two columns. The column on the left indicates you have counted beyond nine. The value 1–9 that you put in the left-hand column will indicate how many times you have counted beyond nine.

It is easier to explain by saying the value in the first column will be 0–9. This will be called the *ones column*. Sometimes the ones column is also called units column. The column to the left of the ones column is called the *tens column* (see Fig. 3-1). The first time you count past 9 you would put a 1 in tens column and start the value in the ones column back to 0. You have always figured that 10 comes after 9. Now we will think of 10 as actually being two digits, with 1 being in the tens column and 0 in the ones column.

When you count beyond 9 the second time, you actually are counting beyond 19. This time you place a 2 in the tens column and again put the lowest value, 0, in the ones column. The number is now 20 (or two, zero). The process is repeated every time you count to the highest value in the ones column. For instance, 30 comes after 29, and 60 comes after 59, and so on. In each case, when you reached the highest value in the ones column, you added 1 to the tens column and started the ones column at the lowest value.

The highest value you can count for two digits is 99. You use the same theory to count beyond 99 as you did every time the ones column came to 9. In this case, you add one to the next column, which is the hundreds column, and return the other two columns, tens and ones, to their lowest value, which is 0. Now you have the next number larger than 99, which is 100 (or one, zero, zero). You call this number, one hundred. When the tens and ones columns become full (199), again you add one more value to the hundreds column to reach 200. This process is repeated until you reach 999. At this time the same rules are used: you add one to a new column to the left of the hundreds column. This new column is called thousands. When the thousands column becomes full, the next column is ten thousands, and the next column is hundred thousands.

You should begin to see that, as each column is added, it is exactly ten times larger than the column before it. In other words, the column to the left of the tens column is the hundreds column because $10 \times 10 = 100$. The column next to the hundreds column is the thousands column because $100 \times 10 = 1000$. The column beyond thousands is ten thousands because $1000 \times 10 = 10,000$. No matter how high you want to count in the decimal system, the next column to the left will be ten times larger than the preceding column.

Let's examine several numbers in the decimal system and show their value. For example, 764 has a 7 in the hundreds column, 6 in the tens column and 4 in the ones column. You could actually show the number's value by multiplying 7 by 100 to get 700, multiplying 6 by 10 to get 60, and 4 by 1 to get 4. When you add these values, you have a total of 764 (see Fig. 3-2).

Figure 3-1 Column weights for decimal numbering system.

Thousands   Hundreds   Tens   Ones

```
                            ( 7 × 100  = 700
Hundreds Tens Ones          { 6 × 10   =  60
    7      6    4            ) 4 × 1    = + 4
                            (            _____
                                          764
```

Figure 3-2 Decimal numbering using columns for weights.

You may wonder why we would go to all this trouble just to count numbers. The main idea is to show the numbering system you are most familiar with, then to look at the other systems. Binary, octal and hexadecimal systems all work the same way.

BINARY NUMBERING SYSTEM

The binary numbering system is the common numbering system used for computer operation. The binary system is called the *base two system*. This means the binary system has just two values. These values are 1 and 0. Since they are the only two values in the system, columns must be used to show position, or weighting, to count beyond these values. The best way to understand the binary system is to count to 15.

Figure 3-3 shows the binary and decimal system counting up to 15. From this example you can see that the first two values for both numbering systems are the same. The binary system must use a second column once the values go beyond 1. Actually, the binary system operates very similar to the decimal numbering system. Since 1 is the highest value in the binary system, you must add a column to the left of the first column to count beyond 1. This new column is called the *twos column*. Since the ones column is full, you must add 1 to the twos column and put the lowest value in the ones column. This means the binary value for 2 is 10 (one, zero). This actually shows a 1 in the twos column and 0 in the ones

			Binary		
		Eights	Fours	Twos	Ones
Decimal	0	0	0	0	0
	1	0	0	0	1
	2	0	0	1	0
	3	0	0	1	1
	4	0	1	0	0
	5	0	1	0	1
	6	0	1	1	0
	7	0	1	1	1
	8	1	0	0	0
	9	1	0	0	1
	10	1	0	1	0
	11	1	0	1	1
	12	1	1	0	0
	13	1	1	0	1
	14	1	1	1	0
	15	1	1	1	1

Figure 3-3 Binary and decimal numbers.

column. The binary value for 3 is 11 (one, one). This is 1 in the twos column and 1 in the ones column. Now that both columns are full, or have the highest value in them, a column must be added to count beyond 3. This new column is called the *fours column*. The binary value for 4 is 100 (one, zero, zero). This represents 1 in the fours column, 0 in the twos column, and 0 in the ones column. You can see that 5 is 101 (one, zero, one), which is 1 in the fours column, 0 in the twos column, and 1 in the ones column. Another way to say this is that 5 is made up of one 4 and one 1, or 4 + 1 = 5.

The next columns in the binary system are eights, sixteens, thirty-twos, and sixty-fours. Again, you can see each new column is found by multiplying the previous column by the number of digits in the numbering system. Since the binary system has two digits, each column is multiplied by 2 to get the value of the next column.

Another point to remember is that you must identify which numbering system you are working in. Since 10 can represent ten in the decimal system and 2 in the binary system, a subscript is used to identify what numbering system is being used. A small 2 is used to show you are using the binary system. If you are using the decimal system the subscript value is a 10.

Generally, the subscript for the decimal numbering system is omitted because this system is the most used in daily life. It is very important to use the subscript when you are working around P.C.'s or computers.

The reason the binary system is used so often in digital electronic systems, such as programmable controllers, is that the two values in the system, 1 and 0, represent the two possible electrical conditions of a switch. The switch could be on, or it could be off. In most electrical systems, the "on" condition is designated as a 1 and the "off" condition is designated as a 0.

This means the computers and processor chips inside programmable controllers can handle numbers and data because they have virtually thousands of switch-type circuits. By designating each switch a weighted value, or column, such as ones, twos, fours, eights, or sixteens, the processor can handle very large values. To understand the binary numbering system, you will need to practice converting binary values to decimal and decimal values to binary, listed at the end of this chapter.

BINARY-CODED DECIMAL SYSTEM

The binary-coded decimal numbering system, also called BCD, is a combination of the binary and the decimal systems. The system works basically like this: Binary values are the only values used in the system. This means only 1s and 0s are allowed. The weighting of the binary values are traditional binary weights for the first four values. This means the ones, twos, fours, and eights columns are used. Each group of four binary columns allows you to count binary from 0 through 9. This should sound familiar since they are the values of the decimal system. Another

100s   10s   1s   **Figure 3-4** Example of BCD weighted
0000   0000   0000 columns.

group of four binary digits are used together for the tens column, enabling you to
count values 10 through 19. These four digits are grouped to the left of the first
group of four binary digits. These four digits represent the tens column in the
decimal system. If you need to count beyond 99, another four binary digits are
grouped to give you the hundreds column. An example of BCD weighted columns
is shown in Fig. 3-4 and Fig. 3-5.

The BCD system is a very usable number system because it can communicate
in binary with the computer and it is coded into decimal values for you.

The BCD system is generally discussed in terms of the number of decimal
units the BCD represents. In other words, a one-decimal digit BCD needs 4 binary
digits, or bits. A two-decimal unit BCD number needs 8 binary digits, or bits, and
a three-decimal unit BCD number needs 12 bits. The BCD value needs more bits
to operate than the binary system, but it is easier to read. If the computer is
communicating with its memory, or modules, it will probably use the binary system.
If the P.C. is communicating with its CRT, or data display, it may need to convert
its numbers to BCD so humans can read the values in decimal form. The most
common devices that use BCD numbering system are thumbwheels and 7-segment
LED displays. Several problems are presented at the end of this chapter for you to
convert from binary to BCD values and from decimal to BCD values. These problems will provide you with practice for using these systems. You may need to
reread this part of chapter 3 if you have problems converting these values. It is
essential that you fully understand them and are capable of converting numbers
from binary to BCD and from BCD to decimal. If you still have trouble understanding these systems be sure to ask for help.

	Tens	Ones		Tens	Ones
1	0000	0001	11	0001	0001
2	0000	0010	12	0001	0010
3	0000	0011	13	0001	0011
4	0000	0100	14	0001	0100
5	0000	0101	15	0001	0101
6	0000	0110	16	0001	0110
7	0000	0111	17	0001	0111
8	0000	1000	18	0001	1000
9	0000	1001	19	0001	1001
10	0001	0000	20	0010	0000

Figure 3-5 BCD count to 20.

OCTAL NUMBERING SYSTEMS

The octal numbering system is another numbering system commonly used in P.C.'s
and computers. The octal system is also called the *base eight system*. The octal
system has eight values. These values are 0–7. To count to values higher than 7,
you must add columns just as you did in the decimal and binary system. The

Figure 3-6 Octal numbering system.   64s   8s   1s

columns for the octal numbering system are shown in Fig. 3-6. These columns are determined just like columns in other numbering systems, by multiplying each column by the number of values in the system. This means that each column is multiplied by 8 to get the value of the next column. The columns in the octal numbering system are ones, eights, sixty-fours, etc. Figure 3-7 shows the use of the octal system to count to 64. As you will notice, the values 8 and 9 will not show up in the octal numbering system.

The octal numbering system is commonly used in P.C. systems, especially for identifying I/O modules. It is useful because most processors work with 8-bit I/O structure. In this way, the processor can use one I/O line for each octal digit. You must also remember that if you use the octal numbering system you must use a small subscript 8 to identify the system.

It should also be pointed out at this time that the octal numbering system is not especially suited for conversion to other systems. Instead, it is meant to be used on its own. This means that you simply use the octal system and don't worry what it converts to. If the values 0–7 or 10–17 are used as addresses for modules, just use them, and don't try to convert them to decimal or binary. You'll soon begin to understand how the system works as you use it for module addresses. You will also find it is quite useful because each data memory word is either 8 bits or 16 bits long, and the octal numbering system represents these values well.

Decimal	Octal	Decimal	Octal	Decimal	Octal
0	0	26	32	51	63
1	1	27	33	52	64
2	2	28	34	53	65
3	3	29	35	54	66
4	4	30	36	55	67
5	5	31	37	56	70
6	6	32	40	57	71
7	7	33	41	58	72
8	10	34	42	59	73
9	11	35	43	60	74
10	12	36	44	61	75
11	13	37	45	62	76
12	14	38	46	63	77
13	15	39	47	64	100
14	16	40	50		
15	17	41	51		
16	20	42	52		
17	21	43	53		
18	22	44	54		
19	23	45	55		
10	24	46	56		
21	25	47	57		
22	26	48	60		
23	27	49	61		
24	30	50	62		
25	31				

Figure 3-7 Decimal to octal conversion.

INTEGERS, FLOATING POINT NUMBERS, AND HIGH-ORDER NUMBERS

When the decimal numbering system is used in a P.C., there are several files, or registers, that may be used to help with the special numbers that might be encountered. These special numbers and files are called integers, floating point and high-order numbers.

Integers are simply all the whole numbers used in the decimal system. This means that only whole numbers and no fractions or decimal points are used. The integers, or whole numbers, can be both positive and negative numbers. The maximum value of both positive and negative numbers that can be used in the P.C. will depend on the type of processor chip.

Floating point numbers used in the P.C. will allow the use of a decimal point in the system. For example, numbers like 6.48 and 3.26 are allowed in a floating point file. Usually the processor will determine a standard placement for the decimal point, such as two or three places.

You may wonder why the fuss is made about integers and floating point numbers. Essentially, the processor can operate much faster and use less memory when only integers or whole numbers are used. If the values are rather large, the decimal point is generally unnecessary. Any time a decimal point is used, more memory is required. If the numbers are rather small and a great deal of detail is needed, the decimal point will be necessary. At this time, the extra memory and loss of speed will be worthwhile to attain the needed accuracy.

Some P.C.'s allow only integers, while others, like the Allen-Bradley PLC-3, Texas Instruments Model 560/565, and the Modicon 584, enable the user to use both integers and floating point values.

High-order numbers and high-order files are simply a way to treat very large integers. These values are usually grouped separately in the processor memory because high-order integers require larger registers. Usually the division between normal integers and high-order integers takes place at the value 32,767. Any number larger than this value is considered a high-order integer. Any value lower than this number is considered a normal integer. If the value is negative, the high-order integers start at $-32,768$. This value is chosen because of the 16-bit processor. If you calculate 2 to the 15th power, you can see the value is 32,768. The reason the 15th power is used instead of the 16th is because one placement is generally saved, or reserved, for the sign $+$ or $-$. This is called the sign bit. Figure 3-8 shows an example of these high-order numbers. From this example, you can see that the values to the positive side take up 15 bits and the values to the negative side take up 15 bits. Zero is considered the first value on the positive side.

High-order numbers are needed in some complex P.C. systems to keep track of very large numbers. Registers used to store these large numbers, such as answers to multiplication functions, are usually 16 bits. These larger registers are usually grouped in a section of processor memory called the *high-order integers file*. The high-order numbers are needed for very large numbers, but you must understand

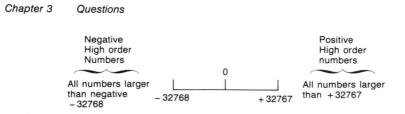

Figure 3-8 High order numbers.

that they take up larger portions of memory and slow the processor when they are used in calculations. Some P.C. systems do not provide high-order integer numbers and files. Instead they connect several normal registers together to form one large register. This is known as a *double precision register*. Double precision registers can handle very large numbers, but they are also slow and take more memory.

When you use any of these numbering systems you will have to be cautious, especially when you are converting from one system to another. After you read the other chapters in this text, you will begin to see the need for all these numbering systems. If you come across an example of a number system that is a little difficult to understand, simply return to this chapter for a quick review. Be sure to try the problems and questions at the end of this chapter before you proceed to chapter 4.

QUESTIONS

1. Explain why weighted columns must be used to count beyond 9 in the decimal system.
2. List the ten integers that are used in the decimal numbering system.
3. List the two permissible integers in the binary system.
4. Show the binary numbers for decimal numbers 0–15.
5. Name the columns used for weights in the binary numbering system.
6. Explain how the column weights of ones, twos, fours, eights, and sixteens were determined for the binary system.
7. Explain why the binary number is so usable in P.C. and computer systems.
8. Explain how values such as 10 and 11 in the decimal system are not confused with values 10 and 11 in the binary system.
9. Explain how the binary-coded decimal numbering system operates.
10. Show the BCD value for the following decimal values: 6, 23, 48, 123, 298.
11. Explain why the BCD numbering system is used so often in P.C. systems.
12. Name one input and one output device ideally suited for the BCD numbering system.
13. How many binary digits are needed for a BCD value of 999?
14. Name one drawback in using the BCD system instead of the binary system.
15. Show the octal values that represent the decimal values 0–15.
16. Explain why the octal system is generally used to number output and input modules.
17. Explain why the octal system is usually not converted to decimal or binary values.
18. Convert the decimal value 136 to binary.
19. Convert the decimal value 136 to BCD.

20. Convert the following binary values to decimal.
    1010110
    100110
    1110110
    000110

21. Convert the following BCD values to decimal values.

    1000   1001
    0010   0111
    0101   1001
    1001   0000
    0100   0001

22. In reference to problems 20 and 21, which system was easiest to convert? Why?

chapter 4
Programming Panels

The part of the programmable controller system that you will come in contact with most often is the programming panel. Figure 4-1 shows a picture of a typical programming panel. In this chapter, you will see that each manufacturer has designed their programming panel for specific use in the P.C. system. Another point you will notice is that most P.C. manufacturers provide several programming panels that fit the application, abilities, and costs of the total P.C. system. In all systems, the programming panel is the interface between the P.C. and programmer, or the troubleshooting technician.

The programming panel provides an area to view the program. This may be a cathode-ray tube (CRT) (which is like a television screen), a 7-segment type display, or a liquid crystal display (LCD). The manufacturer provides a booklet for general operating instructions. This chapter provides some insight and explanation about what you should expect to do with the programming panel. You should notice that each programming panel may have hundreds of keys and functions. You should understand that you are not expected to know the function of each key the minute you sit down to program. Instead, you should learn their functions as you use them. Obviously, the ones you use the most will be memorized. The other functions, which are rarely used or complex, will require the use of the instruction manual each time you program them.

Figure 4-1 A typical programming panel. Picture courtesy of Gould Modicon Programmable Controller Division.

BASIC TYPES OF PROGRAMMING PANELS

Each P.C. manufacturer has a standard programming panel. Figures 4-2, 4-3, 4-4, and 4-5 show pictures of typical programming panels. You will notice that Allen-Bradley calls its panel an industrial terminal; General Electric calls its panel Program Development Terminal (PDT); Texas Instruments calls its programming panel Video Programming Unit (VPU); and Square D calls its programming panel CRT Programmer. In each case, the unit is the device used to enter programs into the P.C. processor through a keyboard.

Each programming panel has a keyboard, which allows the user to enter programs into the processor. The keyboard is usually divided into several sections. Figure 4-6 shows the typical division of the keyboard for the Modicon P180 programming panel. You can see the keyboard is divided into four sections. These sections group together keys with similar functions to make their operation easier.

The first group of keys is called the *cursor control group*. The cursor is the little white, or lighted, square that appears on the CRT. The cursor will show you where you are in the program and display the point where symbols will be entered. The cursor control keys also help you to move around the program page that shows on the CRT. These keys are arrow keys and will move the cursor block around the screen. The other cursor control keys, GET NEXT and GET PREV, allow you to move ahead to the next page of the program or behind to the previous page. This

Figure 4-2 Gould P190 programming panel. Picture courtesy of Gould Modicon Programmable Controller Division.

Figure 4-3 Allen-Bradley programming panel. Picture courtesy of Allen-Bradley Company.

Figure 4-4 Square D programming panels. Picture courtesy of Square D Company.

Figure 4-5 Texas Instruments programming peripherals. Picture courtesy of Texas Instruments.

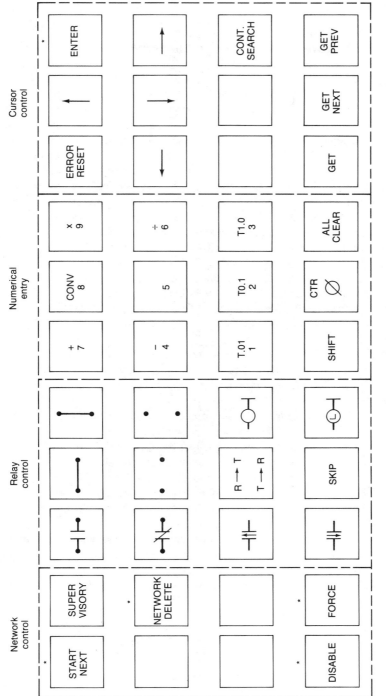

Figure 4-6 Gould P180 programming panel keyboard. Diagram courtesy of Gould Modicon Programmable Controller Division.

allows you to move from program page to program page. The GET key provides a very unique function for Modicon P.C.'s. The GET key allows the user to check the contents of a register or the status of a specific input or output. When the cursor is moved to the lower right section of the CRT, the user can key in the input, output, or register number and push the GET key, and the input, output, or register number will appear under the cursor along with its status, or current value. If the GET key is pushed again, the next consecutive register and its contents will be displayed. This can be continued to inspect the values in consecutive registers.

The CRT is divided into several areas. The top portion of the screen displays the present program. The lower portion of the screen has four status areas. Figure 4-7 shows these four status areas. The assembly area is the first area on the left side. When you touch a symbol key on the keyboard, the symbol or number will appear in the assembly area. This area provides a location to construct logic and inspect it prior to entering it into the program. When the ENTER key is pressed, the symbols in the assembly area are entered into the program.

The next area in the status area is the message area, where error messages are displayed. The machine area is to the right of the message area. The type of processor and size of memory are displayed here. The area on the far right of the status area is called the *reference area*. The reference area provides room for displaying register data and I/O status. When the GET function is used, the register data or I/O status is displayed here.

The SEARCH key is another cursor control key that allows the user to move the cursor to display a specific device in the program. This function helps the user move directly through the program to a specific device for inspection, editing, or troubleshooting.

The ENTER key allows the user to enter the device that is keyed into the assembly area of the CRT. This allows the user to assemble a program device on the CRT and look at it prior to entering it into the program. As long as the device is in the assembly area of the CRT, it can be changed, altered, or written over before it is entered into the program memory. Once the ENTER key is used, the device will go into the program. If changes must be made after a device has been entered into the program, the cursor can be moved over the device, and the changes can be written over the original program.

The next group of keys is the numerical enter keys. These keys are dual function keys. This means that each key has two functions, an uppercase function and a lowercase function. When you press the key normally, the lowercase function is selected. To select an uppercase function, use the SHIFT key prior to touching

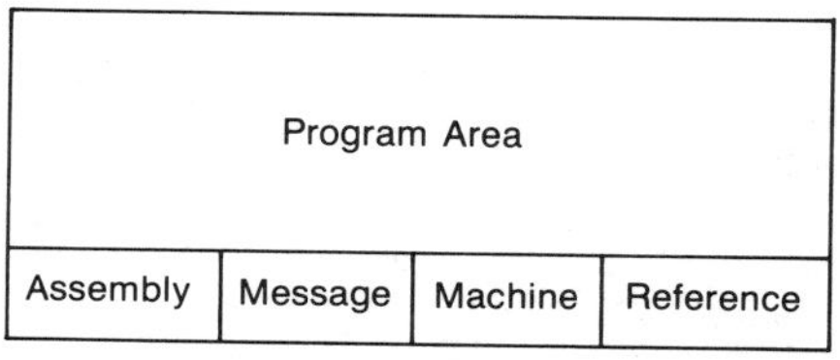

Program Area			
Assembly	Message	Machine	Reference

Figure 4-7 Layout of CRT screen and status area of P180 programming panel.

the numeric key. The CRT will indicate when the keyboard is in the SHIFT mode. The ALL CLEAR key is used to clear the assembly area of the CRT. This key will not alter any part of the program that is in memory.

The next group of keys includes the relay control keys. These keys include normally open and normally closed contacts, output coils and latch coils, transition contacts for one-shot devices, and vertical and horizontal shorts and opens. As you begin to program you will get a better understanding of their functions.

The last group of keys is the network controls. These keys include the NET-WORK DELETE key. The lowercase function of this key allows the user to delete any single device in the program. The Modicon 484 format requires that the device to be deleted is to be the last device in a program line and in the last line on that page. The uppercase function for this key allows the user to delete a complete program page at one time. This allows the user to remove unwanted program lines.

The disable and force function keys allow the user to disable and force outputs. You should not use these two keys if the P.C. is connected to a machine until you fully understand how they may cause uncontrolled machine motion that may endanger personnel and the machine.

The START NEXT key will start a new, clean program network, or page. Each time you push this key, a new page appears on the CRT, and the page number increases by one. The SUPERVISORY function key allows the user to put the processor into one of six modes. When the SUPERVISORY function key is depressed, a menu of six supervisory functions is displayed (see Fig. 4-8). Since all of these supervisory functions are important, they must be followed by a 7 to confirm your choice. If 1 is selected the processor is placed in the STOP mode. This allows the user to continue programming, but input and output signals will not function. When 2 is selected, the processor is placed in RUN mode. In the RUN mode the processor will continually scan the program as it receives input signals and energizes outputs. If 3 is selected, the memory will be totally cleared. It must be noted here that you should not use the CLEAR MEMORY key unless you have a copy of your program on tape or you wish to remove all programming from the processor. If function 4, 5, or 6 is selected, the processor will load a program into memory from tape, dump a program to tape from memory, or verify a tape against memory. Each of these functions requires some storage device, such as a tape recorder, to be connected to the ASCII port.

As you have noticed, this is the keyboard for a Modicon P180 programming terminal. Other manufacturer's programming panels have similar groups of keys that provide the same types of functions. Some functions on all programming panels

Figure 4-8 Supervisory functions for Modicon 484.

0: Exit Supervisory State
1: Stop Controller
2: Start Controller
3: Clear Controller Memory
4: Load Memory
5: Dump Memory
6: Verify Memory
7: Confirm

will only operate when the MEMORY PROTECT key is in the off position. If memory protect is on, the program is protected against changes of any kind. MEMORY PROTECT is usually activated with a key either on the processor or the programming panel.

Other types of programming panels, such as the Modicon P190, Texas Instruments VPU200, and the G.E. PDT, provide a method for loading and storing programs right at the programming panel. Figure 4-9 shows a picture of the tape loader feature of the Modicon P190. As you can see in this picture, a small cassette tape can be inserted into the tape player. These tapes are used to configure the P190 as a tape loader for dumping and loading programs, or as a keyboard to program the Micro 84, 484, 584, or 884. Each of these systems requires its own tape. In essence, the P190 is a "dumb" terminal until the programmer places a tape into it to tell it what device you want it to become. Since it can perform each of these jobs when the proper tape is loaded, the P190 is very versatile.

The Texas Instruments VPU200 has a disc drive built into its front panel. The disc is used to configure the T.I. system and store programs just like the

Figure 4-9 Gould P190 tape loader function. Picture courtesy of Gould Modicon Programmable Controller Division.

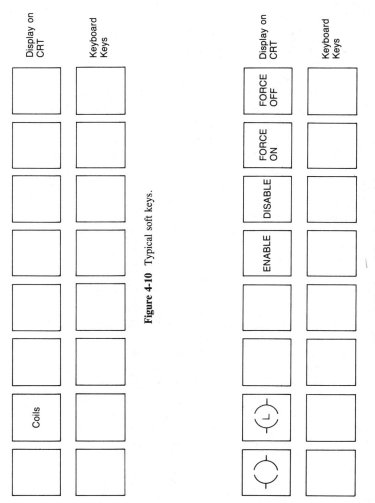

Figure 4-10 Typical soft keys.

Figure 4-11 Second function for soft keys.

Modicon P190. The G.E. PDT may have the tape recorder built-in as an option. The recorder is used to load or store programs from cassette tapes.

Each of these systems provides features beyond simple programming. You must remember that in some machines the tape or disc device is an option and can be disregarded until you want to store a program. In other systems, such as the Modicon P190, it is an integral part of the programming panel that must be used to get the programming panel to operate correctly. In this case, you should seek some help as you use the P190 tapes for the first time. It is not a difficult system, but it does add some confusion to a first-time user.

Another feature that some systems provide is multiple key functions for a single key. The Allen-Bradley system allows the use of a series of overlays that snap onto the keyboard. In other words, the keyboard cover can be changed, giving each key a new function. These overlays enable you to use the T3 and T4 industrial terminals for programming ladder diagrams, alphanumeric (letters and numbers like on a typewriter), or as a graphic keyboard. This function allows a wide variety of uses for the same terminal.

Another type of multifunction keyboard is the use of "soft keys." Soft keys are a set of programmable or function selectable keys. For example, the Modicon P190 programming panel provides eight soft keys. Figure 4-10 shows how these keys are positioned on the panel and CRT. You should notice that every other key is colored light and then dark. The light and dark blocks on the CRT display the function of each of the corresponding light and dark keys. From Fig. 4-10 you can se that if the second key is selected, its function at this time is coils. When the key is selected, the names or functions of the soft keys change to types of coils.

Figure 4-11 shows the six new selections, or functions, the soft keys now represent. In this case, if you pushed the second key again, you would select a latch coil (L) to be entered into the program. As you can see, each soft key may have dozens of functions depending on the previous selection. This allows the user to have virtually hundreds of functions without having hundreds of keys spread out on a very large keyboard.

Many other programming panels also use soft keys or multiple function keys. Some units use multiple menu and numerical keys to provide multiple functions. Sometimes these systems seem very difficult to understand at first. If you remember to watch the CRT, it will usually remind you of what function the multi-function key represents at any given time.

KEYPADS

Some systems, like the Allen-Bradley PLC-3 and the Modicon 584, provide a keypad on the processor for limited system functions. The picture in Fig. 4-12 shows an example of the 584 keypad and 7-segment display. The keypad is mounted on the front panel of the 584 to allow the operator, or troubleshooting technician, a chance to examine the contents in the data registers and the status of certain parts

Figure 4-12 Using Modicon 584's register access panel. Picture courtesy of Gould Modicon Programmable Controller Division.

of the program. Modicon calls this keypad a *register access panel*, or RAP. This keypad allows these limited functions, even when the industrial terminal of a programming panel is not connected. This enables a quick way to check the system without connecting a programming panel. You must keep in mind, some large companies will have dozens of processors and systems operating at one time and only have one or two programming panels available. The keypad provides a means of checking the system without connecting the programming panel to the processor each time.

By now you can see a wide variety of uses for programming panels. Each P.C. manufacturer has a specific function in mind when their system is designed. You will become quite familiar with the operation of each programming panel you use. You will find things you like and don't like if you use several manufacturers' systems. You must remember, though, that your terminal will only operate on the system it is designed for. With practice, you will find the programming panel is a valuable aid in putting programs into the processor, and it will be a set of eyes to see what is in the system. You can be sure that P.C. manufacturers will update the programming panel in future years to make it more usable and more versatile.

First time P.C. users tend to become overwhelmed by the complexity of the programming panel. If you feel this way you should stop to think that most of these keys provide some kind of time-saving function and at some later date you will be glad they are available. I would also strongly encourage all programming panel users to keep a small pocket notebook handy to write down the functions of keys as you learn them. This way you are not wasting time trying to memorize them. Rather you can look up the keyboard functions and the keystrokes you used to enter the program from the keyboard. Don't feel bad about keeping notes and using them. In fact most successful programmers use many notes, and most P.C. manufacturers provide a pocket notebook with all their programming panel instructions printed in them. If you intend to work around programmable controllers, you can figure on using the programming panel almost daily. In time you will find it is an invaluable part of the P.C. system.

QUESTIONS

1. Explain what a programming panel is.
2. What is a CRT, and why is it used with a programming panel?
3. What is a cursor?
4. Where is the cursor found?
5. How do you move the cursor?
6. Explain how the GET key will function for the Modicon P180.
7. Explain how the SHIFT key operates with dual function keys.
8. Explain what the ENTER key does on the Modicon Programming Panel.
9. Explain why it is very dangerous to use FORCE and DISABLE keys on an operating P.C. system.
10. Explain the function of the Supervisory keys for the Modicon system.
11. How do other P.C. systems present supervisory functions?
12. Explain how the Allen-Bradley programming panels keyboard overlays work.
13. Explain what the tape and disc recorders are used for on the programming panel.
14. Explain what soft keys are and how they work.
15. Why are soft keys used on programming panels?
16. Explain why some larger P.C.'s provide a keypad right on the face of the processor panel.

chapter 5

Coils
and Contacts

All programmable controllers have one common feature. They can all receive input signals and send output signals. This means that they can communicate with the real world. It is also the main feature that makes P.C.'s different from regular computers. P.C.'s are designed especially for sending and receiving I/O signals. The input signals will come from switches or contacts in an industrial system, indicating some condition has changed. These industrial inputs may include push-button start and stop switches, limit switches, and selector switches. The outputs will be sent out into the real world to cause some condition to change. Some typical outputs are indicator lamps, motor starters, and solenoids.

The programmable controller must have a program in its memory to react to when it receives these input signals and sends out output signals. The program symbols for a P.C. input will look like a normally open or normally closed contact used in typical electrical diagrams. These symbols are shown in Fig. 5-1. These contact symbols are typically used by most P.C. systems.

The program symbol for a P.C. output will also look similar to symbols used in typical electrical diagrams. In Fig. 5-2 you can see that the symbol will look similar to a circle or two parentheses close together. Some P.C.'s may use a slightly different symbol for an output. The difference between symbols is usually due to the variation between major P.C. manufacturers' choice of processor chips.

Figure 5-1

Typical PC Normally Open Contact Symbol

Typical PC Normally Closed Contact Symbol

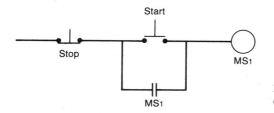

Figure 5-2 Typical P.C. output symbols.

Since all input symbols in a P.C. look exactly alike, a numbering system must be used to tell the switches (contacts) apart. The numbering system serves three purposes. First it is a way to tell contacts apart in the program. Second, the number on the contact in the program will serve as an address for the location of the input module where the real-world switch is attached. Third, the number will serve as a memory address for the contacts in the processor memory.

The easiest way to be introduced to these program symbols is to see a typical electrical diagram of a start-stop switch controlling a motor starter converted to a P.C. program. Figure 5-3 shows this program.

Figure 5-3 Typical electrical diagram converted to a P.C. program.

In Fig. 5-3 you can see that normally open and normally closed push-button switch symbols are used to represent the start and stop switches, and a contacts symbol is used to represent the motor starter auxillary contacts, which are used to seal the normally open start push button. A circle is normally used to represent coils, and this coil is identified as a motor starter by adding the letters *MS* below the symbol. Remember, it is important to use the symbol that represents the switch type, such as a normally open push button, limit switch, or selector switch from a relay or motor starter contact when electrical wiring diagrams are drawn. This indicates to the electrician or troubleshooter the type of switch being used.

In comparison you can see that the P.C. program has only one symbol (—||—) for all switches. This means the start push button, stop push button, and the motor starter seal contacts all have the same symbol. The reason for this is that all programmable controller manufacturers are concerned with the amount of memory space in the processor that each symbol will use. The —||— and —|/|— symbols use far less memory than symbols for push buttons, limit switches, or selector switches. For this reason most P.C. manufacturers use only —||— or —|/|— symbols.

At this point, you should understand that there is a vital relationship between the real-world switch, like the start button, and the contacts that show up on the CRT as part of the P.C. program. You create this relationship when you connect the start button to the terminal on the input module and then use the module identification number, or address, to identify the contact in the program.

Once the contact in the program is given the number, or address, of the input

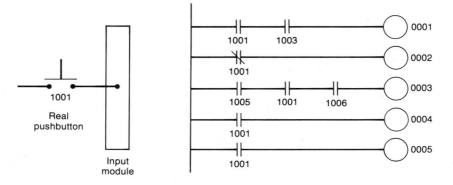

Figure 5-4 One real switch controlling five sets of programmed contacts.

module and terminal, they will reflect the on or off state of the push button. From this point on, whenever the push button is closed, the contacts in the program will change. If an open contact is programmed, it will change to the closed state and pass power through it. If a normally closed contact symbol is used in the program, the programmed contacts will open and stop passing power when the push button switch is closed. This is a very critical relationship between the push button start switch and the programmed contacts.

The unique part of this relationship is that the number used to identify the real-world push button switch can be given to hundreds of other normally open or normally closed contacts in the program. In other words the push button switch that has only one set of electromechanical contacts in the real world can have 10, 15, or 100 different contacts in the program. Figure 5-4 shows an example of the pushbutton having one set of contacts. You should notice that 4 sets of contacts are programmed normally open and 1 set is normally closed. The processor will actuate all five sets of contacts every time the real push-button switch is energized.

You should now see that the P.C. allows one real switch to control hundreds of programmed contacts. The maximum number of contacts the real switch can control will be limited only by the processor's memory size.

Remember, when the real-world contacts change from open to closed, every set of programmed contacts that are controlled by that switch will change. Programmed open contacts will close, and programmed closed contacts will open. You now should see a money-saving feature of the P.C. If your application required a switch with five sets of normally open contacts and three sets of normally closed contacts, you could use a simple, inexpensive single contact selector switch, and it can be programmed on the P.C. as the complex multiple contact switch. You will notice this tactic used in many systems because the single contact switch is much cheaper. Remember, the P.C. can have an unlimited number of programmed contacts controlled by a single real-world switch (up to the limit allowed by memory size).

As memory chips become larger and less expensive, a few manufacturers are experimenting with using other symbols. This will make the identification of programmed contacts a little easier. One important note about the memory in a programmable controller: the larger the memory becomes due to various symbols or increased program size, the slower the program will run. This is known as *scan time*. As the computer takes longer to scan the P.C. program, more errors are introduced into P.C. timer accuracy and the updating of inputs and outputs. This may not mean too much at this point, but it has become the most important feature in most industrial systems used today.

Since the P.C. only uses —||— and —|/|— as input symbols, it becomes a little more difficult to tell a push-button switch from a selector switch or motor starter contact. For this reason, manufacturers have designed numbering systems, or codes, to indicate whether the program symbol represents a real-world switch, such as a push button or selector switch, or if it represents the contacts from a motor starter or other electrical components, such as timers or counters.

Some examples of these manufactured systems will be very helpful in trying to understand P.C. programs. Since there are many different manufacturers of programmable controllers, examples will be presented of the most widely used systems. The idea is to show several methods presently used. From these examples and some help from manufacturer's documentation, you should be able to determine the method your P.C. uses if it is not mentioned here.

ALLEN-BRADLEY NUMBERING SYSTEM

Allen-Bradley has three systems of numbering its contacts and outputs. The first system is for its PLC, and PLC2 families of programmable controllers. The other systems were created for use on their PLC-3 P.C., which is for larger systems and the small PLC-4. From Fig. 5-5, you can see that the scheme for the PLC and PLC2 consists of four groups of letters. The letter *A* indicates whether an input or output will cause the contact to change state. In Fig. 5-5 the possible values (letters or numbers) that *A* could be and what controls the contacts are listed. You can see that if *A* is a 1, the contacts are controlled by a real-world input. If *A* is a 0, the contacts are controlled by an output in the program. (The output can be real, or only exist, in memory.)

From Fig. 5-5 you can also see that two more values *B* and *C* are used. The letter *B* can be a number from 1–7 indicating, the rack number where the module will be mounted. The letter *C* can be a number between 0–7, indicating the module group the module resides in. The letter *D* represents the actual terminal number on the module. The terminal numbers will always be between 0–7 and 10–17. (Re-

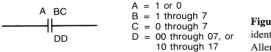

A = 1 or 0
B = 1 through 7
C = 0 through 7
D = 00 through 07, or 10 through 17

Figure 5-5 Allen-Bradley contact identification system. Diagram courtesy of Allen-Bradley Company.

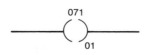

Figure 5-6 Identification number for a
memory coil. Diagram courtesy of Allen-
Bradley Company.

member Allen-Bradley uses an octal numbering system for its rack and modules
identification numbers.)

There may also be times when the contact or coil does not actually come
from a "real" device. This can occur when a coil or output is given an identification
number that uses a rack number which is larger than the system's actual number
of racks. For example, in Fig. 5-6 the number 7 is used as the rack number, but
the system actually has only five racks. Since there is no output module for the
output signal to go to because the system doesn't have that many racks, the output
will be called a *memory output*. This may seem confusing at first, but when you
are looking at the system in industry, you can quickly count the racks and determine
which rack numbers on contacts and coils will be real and which will be memory.

The numbering scheme for the Allen-Bradley PLC-3 is a little different because it is capable of having many more racks and modules than the smaller systems.
In Fig. 5-7, you can see that the letter *A* again represents the control of the contacts.
The letter *A* can be an *I* for input, *O* for output, *T* for timer, or *C* for counter.
Notice that in this system, Allen-Bradley uses *I* and *O* instead of 1 and 0. As in
the previous example, *BBB* represents the rack number and can be values between
000–037 (octal). The letter *C* represents rack number 0–7. The letters *DD* represent
the terminal numbers and can be values 0–7 and 10–17. Again, please note the use
of the octal numbering system.

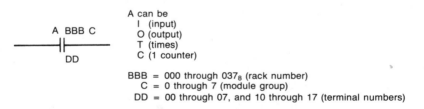

Figure 5-7 Allen-Bradley PLC-3 numbering system. Diagram courtesy of Allen-
Bradley Company.

MODICON NUMBERING SYSTEM

Modicon has its own numbering system for contacts and coils. This system is not
exactly like Allen-Bradley's but it is similar in that it identifies the contacts or input
in regards to the device that controls it and its location in the modules, and it
identifies coils or output as real or memory. Figure 5-8 shows the typical Modicon
numbering system or identification system for contacts in its 484 and Micro 84
processors. In this system, allowable values for "A" are 0, 1, or 2. In this system,
1 indicates that the contact is controlled by a real-world input, such as a limit

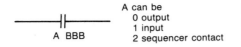

A can be
0 output
1 input
2 sequencer contact

Figure 5-8 Typical Modicon numbering system. Diagram courtesy of Gould Modicon Programmable Controller Division.

switch. The 0 in this system indicates that the contact is controlled by an output that is in the program. The 2 indicates that the contacts are controlled by a sequencer, or drum switch, in the program.

The three numbers represented by *BBB* in Fig. 5-8 indicate the terminal that the input or output is connected to. These numbers start with 001 and run through 256 for the 484, and 01–64 for the Micro 84. This means the numbers for inputs will be 1001 through 1256 or 1001 through 1064. The numbers for outputs will be 0001 through 256 or 0001 through 0064. You should notice that both Allen-Bradley and Modicon use 1 to represent inputs and 0 to represent outputs.

| Module Number (Top of Bottom) | Circuit Number | CHANNEL ONE | | | | | | | |
| | | Housing One | | Housing Two | | Housing Three | | Housing Four | |
		Output	Input	Output	Input	Output	Input	Output	Input
1	1	0001	1001	0033	1033	0065	1065	0097	1097
	2	0002	1002	0034	1034	0066	1066	0098	1098
	3	0003	1003	0035	1035	0067	1067	0099	1099
	4	0004	1004	0036	1036	0068	1068	0100	1100
2	1	0005	1005	0037	1037	0069	1069	0101	1101
	2	0006	1006	0038	1038	0070	1070	0102	1102
	3	0007	1007	0039	1039	0071	1071	0103	1103
	4	0008	1008	0040	1040	0072	1072	0104	1104
3	1	0009	1009	0041	1041	0073	1073	0105	1105
	2	0010	1010	0042	1042	0074	1074	0106	1106
	3	0011	1011	0043	1043	0075	1075	0107	1107
	4	0012	1012	0044	1044	0076	1076	0108	1108
4	1	0013	1013	0045	1045	0077	1077	0109	1109
	2	0014	1014	0046	1046	0078	1078	0110	1110
	3	0015	1015	0047	1047	0079	1079	0111	1111
	4	0016	1016	0048	1048	0080	1080	0112	1112
5	1	0017	1017	0049	1049	0081	1081	0113	1113
	2	0018	1018	0050	1050	0082	1082	0114	1114
	3	0019	1019	0051	1051	0083	1083	0115	1115
	4	0020	1020	0052	1052	0084	1084	0116	1116
6	1	0021	1021	0053	1053	0085	1085	0117	1117
	2	0022	1022	0054	1054	0086	1086	0118	1118
	3	0023	1023	0055	1055	0087	1087	0119	1119
	4	0024	1024	0056	1056	0088	1088	0120	1120
7	1	0025	1025	0057	1057	0089	1089	0121	1121
	2	0026	1026	0058	1058	0090	1090	0122	1122
	3	0027	1027	0059	1059	0091	1091	0123	1123
	4	0028	1028	0060	1060	0092	1092	0124	1124
8	1	0029	1029	0061	1061	0093	1093	0125	1125
	2	0030	1030	0062	1062	0094	1094	0126	1126
	3	0031	1031	0063	1063	0095	1095	0127	1127
	4	0032	1032	0064	1064	0096	1096	0128	1128

Figure 5-9 I/O numbering system for Modicon 484. Diagram courtesy of Gould Modicon Programmable Controller Division.

In larger Modicon systems like the 584 and 884 processors, the number becomes five digits long. The first digit still uses 1 or 0 to indicate inputs and outputs. Five-digit numbers are used because these systems can have more than a thousand inputs or outputs.

You will notice that since Modicon uses sequential decimal numbers to number their input and output modules, they do not need to identify racks and modules. Each module will contain 4 or 8 inputs or outputs, and all you need to do to find a specific input or output is to count by fours or eights. On large systems you may need to use a diagram for I/O, like the one shown in Fig. 5-9. From Fig. 5-9 you can see that the modules are grouped into housings. Each housing will hold a maximum of 32 inputs or outputs in each row. As the rows of modules are racked out, each row will add 32 more inputs and outputs. With a little practice you will be able to find the input and output modules in each housing.

TEXAS INSTRUMENTS INPUTS AND OUTPUTS

Texas Instruments uses a system of identifying their P.C. inputs and outputs similar to the systems just explained. You must remember the contacts and coil symbols will be similar to other P.C.'s. Figure 5-10 shows these symbols. As with other P.C. systems, all the contacts and coils in the program use the same symbols. Since these contacts and coils symbols represent real inputs and outputs, the P.C. must give each symbol in the program a number that will identify its location in the modules and its number in the program. Remember, this is the only way to keep track of the symbols in the program and the inputs and outputs in the system. The P.C. processor needs this numbering system to operate the program, and you, the working technician, need the numbering system to work on the system.

The Texas Instruments numbering system uses the letter X to represent real inputs, such as switches, and the letter Y to indicate that the contacts are controlled by coil, or output. All outputs coils use a Y to indicate it is an output.

Both inputs and outputs are numbered sequentially. This means that the first input will be X1 and each additional input will be X2, X3 . . . in sequential order. Each output is also identified the same way. You should also notice that Texas Instruments, like all other P.C. manufacturers, use these same input and output numbers to identify the modules in their racks, as well as in the program. This allows the technician to use one common identification number for each input, regardless of whether they are looking for the input in the program or in the module. You should also remember this number will also be used by the processor as the address where it stores this contact number in memory. As you know, this is a common practice in all P.C.'s.

Figure 5-10 Typical Texas Instruments' symbols. Diagram courtesy of Texas Instruments.

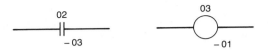

Figure 5-11 Typical contact and coil numbers for Square D. Diagram courtesy of Square D Company.

SQUARE D CONTACTS AND COILS

There are many other manufacturers of P.C. equipment. Each of them use very similar symbols for coils and contacts, yet each seem to develop their own numbering and addressing system. For instance, Square D Company Sy/Max® P.C.'s use two numbers to identify each contact or coil. Figure 5-11 shows a typical coil and contact with numbers for identification. The number above the contacts represents the module number, and the number below the contacts identifies the terminal number. The same format is used for the output. The number above the coil symbol identifies the module number, and the number below the coil symbol identifies the terminal on the module. You should notice that even though this method is somewhat different, it still has some similarities to methods used by other P.C.'s. The point here is that you will undoubtedly find systems that you have never used before. With just a few questions and a little practice, you should be able to figure out any system you run into.

SPECIAL CONTACTS

There are several contacts that have special applications on certain systems. One such contact is called a *transitional* or *one-shot contact*. The symbols for this type contact are shown in Fig. 5-12. The symbol for these contacts shows an arrow pointing up or down. Basically these contacts are different from regular contacts in that they will only stay open or closed in the program for one scan each time they are energized. The set of contacts with the arrow pointing up transitions from off to on when the contact becomes energized. As soon as the processor scans the contacts once and moves on, they return to their normal, off state.

The set of contacts with the arrow pointing down transitions from on to off for one scan when they become energized. This may sound strange, but essentially transition contacts provide one pulse when they are energized instead of staying on continually. This is why they are also called one-shot contacts. These contacts are generally used to show a pulse to a sequencer or counter that help to get past complex timing events. Some examples of these contacts, shown in chapters 8 and 9 advance a sequencer past step zero when it is reset, or control a math function.

Figure 5-12 Transitional contacts.

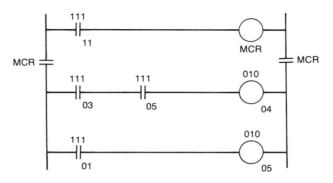

Figure 5-13 MCR contacts.

MCR (MASTER CONTROL RELAY)

Another special type of contact, available on some P.C.'s like Allen-Bradley and Texas Instruments, is called the *master control relay* (MCR).

Figure 5-13 shows a diagram using an MCR. From the diagram, you can see the MCR contacts are the only contacts that can show up in the vertical power rungs.

By using the MCR, you can make zones in the program that can be activated or deactivated for various reasons. This allows the operator or the P.C. to control power to whole zones of an automated system.

One word of caution at this time. *Do not confuse the use of the P.C. MCR with the use of a handwired MCR (master control relay).*

The electrical code requires that every P.C.-controlled system have a hand-wired master control relay wired to a start-stop switch that will provide emergency shutdown to the total system. In the event of any problem with the processor or the P.C. control, an operator or technician can simply hit the stop button and remove power from the P.C. I/O system, which will immediately deenergize all inputs and outputs. This is a safety requirement that must be used with each P.C. system. The MCR available in most P.C. systems is only meant to provide zone control.

You will be using coils and contacts, and outputs and inputs most often in the typical P.C. system. You will begin to gain experience and understand all the possible uses for inputs and outputs. You will also learn to use memory coils to your advantage.

You will also begin to understand the relationship between the input and output devices and the contacts and coils in the program. Once you have gained the knowledge of the operation, troubleshooting and repair of the P.C. system will become much easier.

QUESTIONS

1. Name two purposes the input and output numbering system serves.
2. Why do most P.C.'s use only one symbol for all input in the program?
3. If all inputs in the program have the same symbol, how can you tell them apart?
4. Explain how a switch with one set of real contacts can be programmed to show many sets of contacts.
5. What is the limiting factor to the number of programmed contacts having the same identification number?
6. Explain the systems Allen-Bradley uses to identify its inputs and outputs.
7. What is meant by the term *memory coil*?
8. What is meant by the term *real output*?
9. Explain Modicon's input and output numbering system.
10. Explain Texas Instruments' input and output numbering system.
11. Explain what transitional contacts do.
12. Explain what the P.C.'s MCR contacts can be used for.

chapter 6
Timers

All programmable controllers hae some type of timer available for use in their program. The timers in P.C.'s are very similar to the on delay and off delay electromechanical timers that have been used in industry for many years. This chapter explains the operation and programming of the various timers used in P.C.'s today. Aspects such as time bases, numbers of timers, retentive and nonretentive timers will be explained.

EXAMPLE TIMER CIRCUITS

The easiest way to understand the timers that are found in P.C.'s today is to compare the P.C. timer to the electromechanical timer. The electromechanical timer has two basic parts: the timer clock motor and contacts. The timer's clock is adjustable in increments of 1 second. The clock in Fig. 6-1 is set for 10 seconds. When Switch 1 closes and the clock motor has operated for the 10 seconds, which is the preset time, the timer's contacts change. This means that normally closed contacts would open and deenergize Lamp 2, and the normally open contacts would close and energize Lamp 1. Notice that the timer's clock motor and contacts are identified as TMR 1.

The contacts and timer motor are labeled as TMR 1 so that the person reading the diagram will know which contacts each timer motor controls. Each timer motor may control several sets of normally open and normally closed contacts.

The preset time of the timer can be adjusted mechanically by an electrician. This allows for more or less time delay before the contacts change. These timers

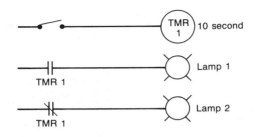

Figure 6-1 Electromechanical timer.

are still used in some electromechanical systems. You can see that since they have a motor, springs, gears, and metal contacts they wear out frequently. They are also more expensive than solid-state timers found on P.C.'s. The basic operation of solid-state timers is very reliable and has been duplicated in the programmable controllers. Since the P.C.'s timers do not have any moving parts, they do not fail as often as electromechanical timers, and their time delay is easy to change through the programming panel.

BASIC P.C. TIMERS

The timer in a P.C. has several basic parts that operate similarly to the electro-mechanical timer, which will make them easy to understand. The P.C. timer also has several functions that can be selected through programming. These functions include: a time base, such as 0.01 seconds, 0.1 seconds or full seconds; off delay or on delay contacts; normally open or normally closed contacts; reset capabilities such as retentive or nonretentive function; adjustable preset time and accumulative times; and done bits.

At first this may sound like a lot of complicated functions just to get a timer to delay the opening or closing of a set of contacts. You will actually find that these functions will allow you to program simple or complex industrial timing activities rather easily. In fact, most electricians and technicians who have used P.C. timers find the electromechanical timers rather restrictive when all adjustments must be accomplished with a tool rather than the keyboard of the programming panel.

TIME BASES AVAILABLE IN TIMERS

One outstanding feature of programmable timers is the variety of time bases available. Most large P.C.'s offer two or three time bases. The most common time bases are 0.01 seconds, 0.1 seconds or 1.0 seconds intervals. The time base selected will depend on how accurate the industrial timing activity must be. For instance, in most general purpose on delay and off delay timers, the 1.0, or whole second, time base is normally selected.

In some industrial processes, such as plastics, the timing activity must be controlled accurately to 0.01 seconds. For this application, the P.C. can time the operation to the nearest 0.01 second and change the P.C. timer's contacts at exactly the right time.

The time base is selected when the program is written and entered into the processor through the programming keyboard. Some manufacturers allow time base selection through time base keys such as 0.01 seconds, 0.1 seconds, or 1.0 seconds, while others require that you type in the 0s and decimal from the keyboard.

Your particular industrial application may require more than one timer. Most P.C.'s will have multiple timers available. Some P.C.'s, such as the Allen Bradley PLC-3, allow up to 10,000 different timers to be used in the same program. You can also select a different time base for each timer used in most P.C.'s.

ON DELAY AND OFF DELAY TIMERS

All programmable timers allow you to delay the time it takes to turn activities on or off. The methods used may look different, but the outcome, or function, is the same. You may find it easier to understand the types of timers whose program looks similar to the motor and contacts diagrams of the electromechanical timers. This type of timer is found in the Allen-Bradley PLC, PLC 2/15, PLC 2/20, and PLC 2/30 processors. Other manufacturers use a different type of program technique called a *function block* to program timers. The function block will be explained later.

The on delay or off delay timer function is selected through the programming keyboard. Allen-Bradley uses an instruction or key called TON for the on delay timer and TOF for the off delay timer. The instructions TON and TOF even sound similar to timer on and timer off. When the program function or instruction sounds like the activity it is performing, it is called a *mneumonic instruction*. Mneumonics is pronounced like the words *new monics*; the first *M* is silent. These instructions have been used in machine language programming of computers for a long time because they help the computer programmer and user remember how the instruction is supposed to function.

Figure 6-2 shows a program diagram for a basic TON on delay timer. The number 030 above the timer identifies this as the first timer. Allen-Bradley locates their timers and counters in the first memory locations after the last rack number. In this case the system has 2 racks and the first timers are located at memory location 030. The TON mneumonic indicates that this is an on delay timer. The timer's time base is indicated by the value 1.0 seconds. The amount of time the timer will run before it times out and changes its contact from open to closed is called *preset time*. The preset time is indicated as 10 seconds for this timer.

This timer will begin to time when contacts 111/01 close and pass power to the clock portion of timer 030. After 10 seconds, the contacts marked 030/15 will change from open to closed and pass power to the output marked Lamp. Bit 15 is controlled by the timer and is called the *done bit* because it changes state when the

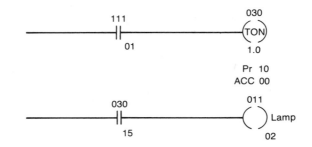

Figure 6-2 Basic TON timer. Diagram courtesy of Allen-Bradley Company.

timer is done timing. You should notice that this operation is nearly identical to the electromechanical timer.

PRESET TIME AND ACCUMULATIVE TIME

The preset time is determined by the programmer and can be selected for any value up to 999 for most processors. The specifications for your P.C. will indicate the largest preset time your processor can accommodate.

When the contacts marked 111/01 close, they will pass power to the timer. Most manufacturers refer to this as *enabling* the timer. Once the timer is enabled, it actually counts in increments that are designated in the time base. In Fig. 6-2 the timer will count up to 10 in precise 1-second intervals. As the timer is counting the seconds from 0 to 10, it is said to be accumulating time. The accumulative time generally starts at 0 and counts up to the value designated in the preset time. Most processors have some means for the programmer or technician to view the accumulative time. This enables the user to see how long it will be before the timer times out. An explanation of how this is accomplished will be covered later in this chapter.

The accumulative time and preset time will be specified by the programmer for each timer in the program. Remember, the preset time indicates how long you want the timer to operate. The accumulative time is the actual amount of time that this timer has accumulated since it has been enabled. When the value in the accumulated time reaches the same value as the preset time, the timer stops timing, even though it is still enabled. When accumulative time equals preset time, all the open contacts that the timer controls will close and all closed contacts the timer controls will open. To get the timer to start timing again, it must be reset.

RESETTING TIMERS

All timers in P.C.'s can be reset to start timing the next event. The TON timer in Fig. 6-2 is reset by opening the enable contacts that are marked 111/01. They could actually be a limit switch, push button, or some type of selector switch in the industrial process. This means that any time these contacts open, timer 030 accumulative value returns to zero. When the accumulative value of a timer is 0, it is said to be in the reset condition. When the contacts close again, the timer will begin timing and the accumulative value will increase until the accumulative value equals the preset time or until the enabled contacts are opened again.

When the timer's accumulative value is reset to 0 each time the enabled contacts are open, it is said to be a nonretentive timer. This term is used because the timer does not retain or remember the accumulative value when the enabled contacts are opened. It simply returns the accumulative value to 0.

There may be times in an industrial process when you want to retain the timer or remember the accumulative value the timer has reached even though the enabled contacts are opened. For instance, you may wish to keep track of the total running time of a machine's motor over a year. The motor may only run three or four hours each day and may turn on and off several times. If you used a nonretentive timer, each time the enable contacts are opened as the motor turned off, the accumulative time would return to 0, and the total running time couldn't accumulate. A retentive timer does not reset its accumulative value to 0 when its enabled contacts are opened. Instead it remembers, or retains, its accumulative value; opening its enabled contacts only stops the timer from running. In order to reset the retentive timer and set its accumulative value to 0, a reset instruction is used. The mneumonic for the retentive timer reset instruction is RTR.

Figure 6-3 shows the program for retentive timer with the RTR reset instruction. Notice the RTR number must match the RTO it is intended to reset.

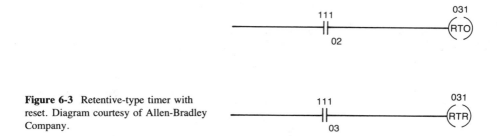

Figure 6-3 Retentive-type timer with reset. Diagram courtesy of Allen-Bradley Company.

Now you can see that the retentive timer can keep track of the accumulative times when the motor is running. Any time the motor is not running, the enabled contacts are opened, and the timer stops. When the motor begins running again, the timer starts accumulating time again. When the motor accumulates enough running time for maintenance, the timer can be reset to record running time until the next maintenance interval.

The retentive and nonretentive timers provide programmers with a variety of timing functions and solutions to many difficult timing problems.

NORMALLY OPEN AND NORMALLY CLOSED CONTACTS

The timers in P.C.'s can control normally open and normally closed contacts. The timer circuit in Fig. 6-2 showed a set of normally open contacts controlled by timer 030. When the accumulative value in the timer equals the preset value, the timer closes its normally open contacts. Allen-Bradley identifies the function of these timer controlled contacts by the number placed below them. In Fig. 6-2 you will notice the number 15 is used; this signifies that this timer contact is controlled by bit 15. Since it will change state after the timer has finished timing, it is called the "done" bit. Other bits the timer can control are bit 17, the enable bit; and bit 16, the time base bit. Bit 17 will be high any time the contacts to the timer are closed. Bit 16 will turn on and off with each pulse of the time base. Each timer in the P.C. can control an unlimited number of these normally open or normally closed contacts. The only real factor limiting the number of timer contacts is the size of the P.C. memory. You should remember memory is the main factor limiting the number of contacts, coils, and timers in a P.C. You generally have more contacts available for program use than you would ever need.

FUNCTION BLOCK TIMERS

Some programmable controllers, like Modicon Micro 84, 484, 584 and Allen-Bradley PLC-3, use a function block to program a timer.

Figure 6-4 shows an example of a function block timer program. This type of timer is programmed in a rectangle called a *function block*. The timer operation is very similar to the electromechanical timer.

The timer in Fig. 6-4 consists of several basic parts that operate very similarly to the P.C. timers discussed earlier in this chapter. There are two sets of contacts used with this timer that control it but are not actually a part of the timer.

The contacts on the top left side of the timer function block are the timer enable contacts. Any time they close and pass power to the top left terminal of the

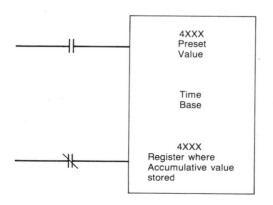

Figure 6-4 Modicon function block-type timer. Diagram courtesy of Gould Modicon Programmable Controller Division.

timer, its clock is enabled, and it starts timing. Any time the enabled contacts are opened and power is removed from this terminal, the timer stops timing.

The contacts on the bottom left side of the timer function block are called *reset*. Any time they open and power is removed from the bottom left terminal, the timer's accumulative value resets to 0. The reset contacts must be closed, or power must be present on the bottom left timer terminal for the timer to start timing.

PRESET VALUES

From Fig. 6-4 you can see the timer has a place for a number on its top line. This value is known as the *preset value*. This value can be any number 0–999 and indicates how many time increments pass before the timer times out. The middle number in the block indicates the time base. Typical time bases are 0.01 seconds, 0.1 seconds, or 1.0 seconds. The 4XXX number at the bottom of the function block indicates the register where the timer is stored. This number may be any value 4001–4256.

The programmer enters the preset value directly from the keyboard. The preset value can be a 4XXX number instead of a value 0–999. If a 4XXX number is used for the preset value, the processor goes to that register and finds the value stored in it. Figure 6-5 shows an example of this type of program. In this program, timer 6 in register 4006 has a number 4012 for a preset value. When the processor finds 4012 as the preset value it goes to register 4012 and gets the value in it. In this case you can see that the value in register 4012, shown in the reference area of the CRT, is 16. This means timer 4006 will have a preset value of 16. You can also see that the value in register 4006, shown in the reference area of the CRT, is 10. This value is called the accumulative value and shows the number of seconds

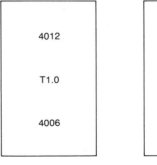

Figure 6-5 Timer and counter in program with register values displayed in the reference area. Diagram courtesy of Gould Modicon Programmable Controller Division.

Reference Area

that have already timed off the preset value of 16. In this case, timer 4006 must have run for 10 seconds at this time.

A very good reason for using a register to store the preset value of a timer is to input this value into a register from a thumbwheel. The thumbwheel could be placed near the machinery being controlled, and preset values could be dialed indirectly to a register, and change the preset value for a timer function block.

ACCUMULATIVE VALUES

The accumulative value in a function block timer will not show up directly in the function block. The accumulative value is the time that has already timed off the timer. This value will be stored in the register indicated by the 4XXX number listed at the bottom of the timer function block. Any register 4001 through 4256 is usable as a storage register. These registers can be used by other functions, so you will need to keep track of which one you have used.

TIMER INPUTS AND OUTPUTS

The timer in Fig. 6-6 operates very similarly to the P.C. timers discussed earlier in this chapter. Two terminals that are part of this timer control its enable and reset functions.

The terminal on the top left side of the timer function block is the timer enable terminal. Any time it is energized and passes power to the timer, its clock is enabled, and it starts timing. Any time the enable terminal is deenergized and power is removed, the timer stops timing.

The terminal on the bottom left side of the timer function block is called *reset*. Any time power is removed from the bottom left terminal, the timer's accumulative value resets to 0. The reset terminal must be closed, or power must be on for the timer to operate when the enable terminal is energized.

The terminal on the top right side of the timer is an output terminal. This terminal receives power the instant the accumulative time value equals the preset time value. By programming an output such as 0001 that is connected to this

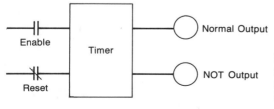

Figure 6-6 Typical enable contacts and reset contact locations. Diagram courtesy of Gould Modicon Programmable Controller Division.

terminal, all contacts controlled by output 0001 become controlled by the timer. Again this type of programming provides the user with an unlimited number of contacts that are controlled by the timer.

The terminal at the bottom right is the *inverted* or "not" output. This means that the terminal on the lower right side of the timer will always be in the opposite condition of the upper right terminal. In other words, any time the upper right terminal is powered, the lower right terminal will not be powered. The right side of the timer function block always provides outputs. The amount of time delay to turn these outputs on or off will be determined by the preset and accumulative times.

TEN-SECOND ON DELAY TIMER

Figure 6-7 shows the function block timer providing a 10-second time delay to an output. In this example, both the enable switch 1005 and the reset 1006 must provide power to the timer for it to begin the time delay. Since reset contact 1006 is programmed normally closed, the timer begins to time at the instant that enable switch 1005 is closed. (Notice that 1005 and 1006 are inputs, and these inputs can be provided by any switch in the industrial process.) You should notice that the preset value for the timer is 10. The time base is selected as 1.0 seconds, and this timer is 4003, or timer number 3.

Switch 1005 must stay closed for the duration of the 10-second time delay. Exactly 10 seconds after 1005 has closed, output 0002 will be energized. This timer has provided a 10-second time delay to output 0002. This timer can be reset to start the time delay again by first opening the enable switch 1005 and then opening and closing the reset switch 1006. The enable switch should then be closed so that the timer starts the time delay again.

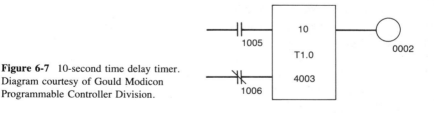

Figure 6-7 10-second time delay timer. Diagram courtesy of Gould Modicon Programmable Controller Division.

TEN-SECOND OFF DELAY TIMER

Figure 6-8 shows a similar timer providing a 10-second off delay to output 0004. You should notice that the timer is controlled by enable switch input 1002 and reset input 1004. The preset value is 10, the time base is 1.0 seconds, and this timer is 4005, or timer number 5.

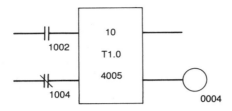

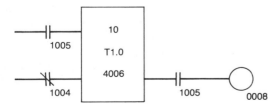

Figure 6-8 10-second off delay timer.
Diagram courtesy of Gould Modicon
Programmable Controller Division.

The lower right output, called the *NOT output* provides power to output 0004 any time the accumulative value does not equal the preset value. This means that as long as the timer has not timed out, output 0004 remains energized. This is sometimes confusing because you might feel that some power must be provided to the timer inputs for the NOT output to power output coil 0004. You should keep in mind that this is a solid-state timer, and any solid-state inverter or NOT logic gate will always have power at its output when the accumulative value does not equal the preset value.

CONTROLLING THE TIMER NOT OUTPUT

In some industrial timing circuits, you may want to provide power to a circuit at the instant the start switch closed, and have the power removed 10 seconds later. If you used the NOT output of the timer to accomplish this, as in Fig. 6-9, power would be provided even before the start switch is closed. Since you want power to be provided to the output only after the start switch is closed, you could put a set of start switch contacts 1005 in series with the NOT output (see Fig. 6-9). In this circuit the reset switch 1004 must remain closed, and switch 1005 must be closed to initiate the 10-second time delay. Remember the NOT output would provide power to the left side of contact 1005. Since 1005 is the start switch, power will not be passed to output 0008 until the start switch contacts 1005 are closed. When the start switch contacts close, the timer starts running and power is allowed to pass to the output 0008. When the timer times out in 10 seconds, the NOT output goes ''low,'' or is deenergized. This action applies power to the output 0008 for 10 seconds and then turns off.

You should notice that there is only one start switch in this industrial process, with two sets of contacts in the program marked 1005. This is one of the strong features of a programmable controller. We can actually program dozens of open and closed contacts marked 1005 that would be controlled by this one start switch.

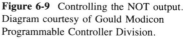

Figure 6-9 Controlling the NOT output.
Diagram courtesy of Gould Modicon
Programmable Controller Division.

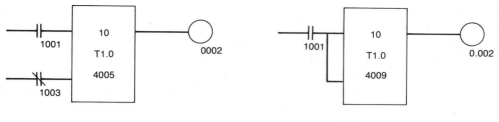

Retentive-Type Timer   Nonretentive Timer

Figure 6-10 Typical retentive- and nonretentive-type timers. Diagram courtesy of
Gould Modicon Programmable Controller Division.

RETENTIVE AND NONRETENTIVE TIMERS

Function block timers can be programmed as either a retentive or a nonretentive
timer. Figure 6-10 shows both types of timers. The function block timers used in
the previous examples have all been retentive timers. This means that the timer
will simply stop timing, but will retain the time accumulated to that point when
the enable contacts are opened and the reset contacts are left closed. If the reset
contacts are opened, the accumulative time is reset to zero, and the accumulated
time is lost.

 If you want to make a nonretentive timer that will reset every time that the
enable contacts open, you would simply use a parallel branch to both the enable
and reset terminals coming out of the enable contacts, as shown in Fig. 6-10. When
the contacts open they deenergize both the enable and reset terminals. By using
the parallel branch from the enable contacts, you ensure that both the enable terminal
and reset terminal on the timer will always see the same signal, a 1 or 0 (high or
low).

AUTOMATIC RESETTING TIMERS

Some industrial timing operations require the time to reset automatically each time
it times out. This timing condition is used quite often in repeating cycles. Figure
6-11 shows the timer in an automatic recycle condition. You can see that the timer
is enabled by contacts 1003. The timer preset is for 5 seconds, and the time base
is 1 second.

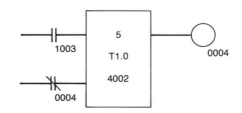

Figure 6-11 Automatic resetting timer.
Diagram courtesy of Gould Modicon
Programmable Controller Division.

When contacts 1003 are closed the timer will begin to time for 5 seconds. At the end of 5 seconds, the timer 4002 times out, and output 0004 is energized. At this instant, the normally closed reset contacts 0004 are opened by output coil 0004. At the very instant the normally closed 0004 reset contacts open, timer 4002 is reset. The reset condition causes the accumulative value to return to 0, and the timer output to coil 0004 is deenergized. When coil 0004 is deenergized it allows its normally closed contacts to return closed. When the 0004 reset contacts are closed, power is again returned to the reset terminal of the timer, and the timer begins timing again. As long as the 1003 enable contacts remained closed, timer 4002 will continue running for 5 seconds, automatically resetting.

You can begin to assume that if a set of output contacts are used to reset a timer, the timer is probably set to automatic reset or cycle. You can also assume that the timer will cycle continuously as long as the enable contacts remain closed to energize the enable terminal. If the enable contacts open, the cycle is disrupted. When the enable contacts are closed again, the automatic resetting cycle will start again. There are many variations of resetting timers. You will have many options available through programming the P.C.

VARIABLE TIME BASES

Some programmable controllers only have 0.1 of a second time base available. In order to get larger values, you will have to use some program variations to attain all timing values. For instance, if you need 4 1/2 seconds time delay, you would program 45 as the preset value and 0.1 seconds as the time base. If you needed a full 10-second time delay you would program 100 as the preset value with 0.1 seconds as the time base. You must remember to divide the preset value by 10 if you use a time base of 0.1 seconds to determine the real time delay in seconds.

In some industrial timing functions you will need a time delay that is larger than the preset value of the timer. For instance, if your timer can only handle maximum three-digit preset values of 999 seconds and you need a time delay of 1500 seconds, it will be necessary to use an extra timer or to add a counter to your timer. From Fig. 6-12 you can see that the first timer 4002 has 999 seconds as a

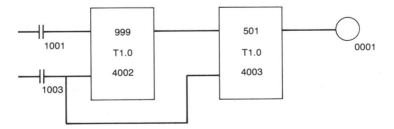

Figure 6-12 Cascading timers. Diagram courtesy of Gould Modicon Programmable Controller Division.

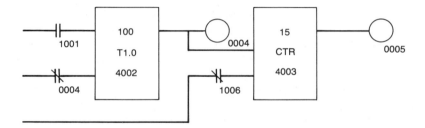

Figure 6-13 Connecting timer to counter to gain extended time delays. Diagram courtesy of Gould Modicon Programmable Controller Division.

preset value. When it times for 999 seconds, it will time out and start the second timer, number 4003. When timer 4003 times for 501 seconds, it will time out and energize output 0001. The total time that has elapsed between closing enable contacts 1001 and energizing output 0001 is the total of both timer preset values (999 + 501 = 1500).

Another way to get time delays beyond the capability of the timer preset limit is shown in Fig. 6-13. In this example, timer 4002 is set for 100 seconds. Every time it times out, it resets itself through contacts 0004. This timer output is also sent as a pulse to the counter (4003). The counter adds one to its count whenever the timer times for 100 seconds. The counter has a preset value of 15. Each time the counter adds a count to its accumulative value, it actually is multiplying the timer's time. For example, if the counter accumulative value is 2, the total elapsed time has been 2 × 100 or 200 seconds. When the counter has counted to 15, it actually is showing a total elapsed time of 100 × 15, or 1500 seconds. At this point, the counter accumulative value equals its preset value, and it sends out output to coil 0005. More information about counters is provided in the next chapter. By combining timers with other timers, or timers with counters, you can actually time values as large as several years. By fully understanding the basic operation of programmable timers you can program nearly any industrial timing function. Several programming problems are provided at the end of this chapter.

TROUBLESHOOTING AND DEBUGGING TIMERS

Sometimes when you try to program or operate a programmed timer, something goes wrong. At first this will bring on a feeling of hopelessness, but it is important to know you really have a powerful troubleshooting aid at your fingertips. The same processor in the programmable controller that takes care of program execution and memory also provides a means of troubleshooting the inputs, outputs, and program. All you need to know is how to use it.

The first place to begin is at the program. Some problems arise when you are trying to enter the timer into the program. The manual for your P.C. shows you a sample format for a timer. You must enter the timer into the P.C. program exactly as shown in the example. The only parts that might be different are numbers on the enable contacts, reset contacts, output, preset time value, incremented time, or the register number where the timer's accumulative value is stored.

When you are entering a timer into the program, watch the CRT; if anything is wrong, the CRT will tell you with an error message. (It is important at this point to remember that the error message is your only means of help. It is not scolding you for doing something wrong; it is merely pointing out something you need to change in order for the processor to accept the timer into the program.) Some common programming mistakes are: having too large a number for the preset value, using register numbers that are already being used or that don't exist in memory, or having the cursor in the wrong place when you are trying to enter the program. Any of these errors will cause the processor to display error messages on the CRT. Read the error message carefully. Sometimes you will need to look in your P.C.'s instruction manual for the exact meaning of the error message, since most error messages are abbreviated. Once you have found the error message, carefully complete the corrective action and reenter the program. If the same error message appears, be sure you are checking the correct error message and corrective action. Many of the error messages are similar, so it is not difficult to get mixed up. You also should not be afraid of changing the program values of the timer in the areas that the error message indicates. Many times a simple change of numbers or values will help. Above all, don't let programming errors scare you. They actually help you learn to program and enable you to reinforce good programming habits. If you still have a problem after several attempts to enter a timer, take a break, talk to someone, and explain what you are doing. You might even consider walking away for a few minutes and having a cup of coffee. You will be surprised when you return to the keyboard that the problem is rather simple. You will also find that once you have found the format error, you will feel rather proud because you have outwitted a very smart device.

In the case of some extra difficult problems, you might try what is called "going back to square one." This means to try entering an example timer program that has worked in the past. Once you get the old reliable program to operate, check to see what you are doing differently, and make the appropriate changes.

One final point, and it is probably the most important point. Computers and programmable controllers are different from other electrical devices in that they very seldom fail on a partial basis. This means that the processor probably will not be the cause of programming problems. When the processor fails, it is usually major, and you can't get anything to happen. Even the keyboard may be dead. This means that 99.9 percent of programming problems are found to be the fault of the person doing the programming. This is not bad, because of the two, it's easier to change a programming attempt than to try to fix the electronics and circuits that are found in the processor.

If you do feel the problem is in the processor or P.C. itself, try to isolate the symptoms and call your local P.C. dealer. Most local P.C. dealers have more technical solutions to try and can continue to provide technical service and parts, if necessary.

There will also be times when the processor has accepted the timer program, but it will not time out or send a signal to an output.

The P.C. system provides several very accurate methods to find the problem. To begin troubleshooting, you should get the CRT, or display terminal, to show the program of the timer. Look closely at the timer, and find the enable and reset contacts. It is a good idea to write their numbers down so you can determine if they are controlled by a real input or memory contacts. Next, check the preset time, time base, and register number. Make sure they are the values that you intended to be programmed. Next you should use the GET function if you are using the Modicon 484 or 584 to display the contents of the timer register. Review the previous part of this chapter if you need help using the GET function or getting the CRT to display the accumulative value in the timer. If you have an Allen-Bradley P.C., the accumulative time will be showing right by the timer. When you can see the accumulative value, check to see if it is less than or equal to the preset value. If the accumulative value is equal to the preset value, the timer must be reset before it starts timing again. If the accumulative value is less than the preset value, the timer has some time left on it before it times out. If the accumulative value is less than the preset value, you are ready to try the timer. Start the timer by closing the switch that controls the enable contacts. (Be sure to check the number on the contacts so that you are using the proper switch.) You can check to be sure that the enable switch is providing power to the timer by looking to see if the enable terminal of the timer is highlighted. This means that the enable terminal on the timer is brighter, indicating it has power.

If you are using an Allen-Bradley timer its symbol will be highlighted. If the timer or the enable terminal does not become highlighted when you turn the enable switch on, you have found the problem. Now you need to check further to find something to repair. The problem will either be somewhere between the real switch used to enable the timer, or in the input/output section if the enable contacts are controlled by a real input. If the contacts are memory contacts, you will need to find the coil that controls these contacts to see what must be closed to energize the coil.

In Fig. 6-14, you can see the diagram for a real switch used as the input. Troubleshoot this using the same process explained in the earlier chapter on inputs and outputs. This basically means you are looking for an input indicator light to glow when the switch is closed. If the indicator lamp does not light when the switch is closed, then the problem must be in the switch, wiring, or power supply. If you do get an indicator lamp to glow, but the CRT will not highlight the contacts in the program, the problem must be in the input module or processor.

If the enable contacts are numbered to indicate that they are controlled by memory contacts, then you must locate the memory coil that causes these contacts to open and close. The best way to find these contacts is to use the search function

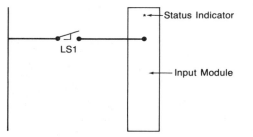

Figure 6-14 Limit switch for input.

that is explained in chapters 1 and 15. If you still have trouble executing the search function, be sure to check your programmable controller's instruction.

Once you have found the memory coil, make sure that its logic rung is true and that it is energized. If you cannot get the memory coil to highlight, then you must look for the cause of the problem in the part of the program that controls the memory coil.

Now that you have the enable terminal or the timer highlighted, its accumulative value should begin timing. If it still does not time, check to be sure the reset terminal is highlighted if it is a function block timer. If it is an Allen-Bradley timer, check to see that the reset coil is not energized. If the reset terminal is not highlighted for the function block timer or the reset coil is energized for other timers, you must troubleshoot their program rungs to find the problem. Remember, for a function block timer to begin timing, it must have its enable terminal and its reset terminals energized. For an Allen-Bradley timer, it must have the timer highlighted and its reset coil unhighlighted, or deenergized.

If the accumulative time does not change when these conditions are met, check to see if the timer needs to be reset. If the timer accumulative value changes but the timer is not keeping accurate time, be sure to see if the timer register is being used by mistake somewhere else in the program.

Now that you have the timer accumulative value operating, make sure that the output is highlighted every time the timer's accumulative value equals its preset value. In the case of the Allen-Bradley timer, its bit 15, the "done" bit, will change from open to closed or from closed to open.

Remember the inverted output on the function block will operate just the opposite of the normal output. You very seldom find a problem where the output terminal or the done bit will not be highlighted when the accumulative value equals preset. But if you do, simply select a different register for the timer. Most output problems will occur where a real output outside the processor will not energize rather than in the memory of the P.C. In this case use the troubleshooting procedures for external outputs discussed in chapter 1. By using these techniques and a little common sense, you should be able to test all timers for operation and find problems or faults wherever they occur.

QUESTIONS

1. Explain the operation of the basic on delay and off delay timer.
2. What is the "done bit" used for?
3. How many contacts can be controlled by the "done bit"?
4. Explain the operation of the timer function block. Include the following:
   (a) enable terminal
   (b) reset terminal
   (c) preset value
   (d) time increment
   (e) accumulative value
   (f) output (on delay)
   (g) NOT output (off delay)
5. Explain the operation of a retentive timer. Include reset operation.
6. Explain the operation of a nonretentive timer. Include reset operation.
7. Explain how you can make a timer automatically reset itself.
8. Explain how you would make a timer show a 65-hour time delay.
9. List the typical time increments available on P.C.'s today.
10. Explain how you could tell how much time a Modicon timer has left to delay (accumulative time).
11. Explain how you could tell the accumulative time on an Allen-Bradley timer.
12. Show a program that will turn off an output after 17 seconds.
13. Show a program that will turn on an output after 25 seconds.
14. Modify the program in problem 13 so that it will only be on for a total of 25 seconds and then turn off.
15. Show a program that will delay an output from turning on for 90 hours.
16. Show a program that will use the preset value found in another register to determine the time delay.

chapter 7
Counters

Counters in programmable controllers have many industrial uses. They are used for such simple operations as counting the number of parts being made by a machine. They are also used in such complex systems as counting very large quantities and keeping track of multiple events by cascading several counters. P.C. counters also have the ability to control an unlimited number of contacts in a program just like timers. These abilities make the P.C. counter very useful to programmers. This chapter helps you understand the operation of P.C. counters and gives you some example programs that show their many users. A comparison of P.C. counters to the electromechanical counter will help you better understand the P.C. counter. The electromechanical counter uses springs, cams, gears, and solenoids to advance a cylinder with numbers on it to keep track of industrial events that are being programmed. In most processors, counters and timers share the same registers or data bytes. This means that if your processor has room for twenty timers and counters, the total number of counters will depend on how many timers and other registers you have programmed.

P.C. COUNTER OPERATION

Counters are programmed similar to timers. Figure 7-1 shows a typical counter program with normally open contacts 110/03 as the enable contacts for the counter. The enable contacts work similarly to timer enable contacts in that they control the counter, but the counter only operates when the enable contacts transition from off

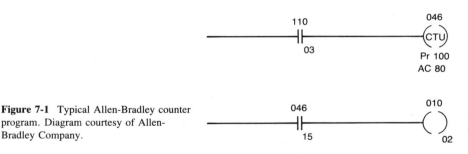

Figure 7-1 Typical Allen-Bradley counter program. Diagram courtesy of Allen-Bradley Company.

to on. This means that 1 is added to the counter every time the enable contacts go from open to closed.

The preset and accumulative values in the counter are exactly like those in the timer. This means that you can place a number in the preset value up to the maximum value allowed by the processor, and when the accumulative value equals the preset value, the counter *done bit*, bit 15, changes from open to closed, or off to on. The done bit can then be used in the program as many times as needed to cause some type of output.

Programmable controllers also have the ability to control a number of programmed counters. The exact number of counters will usually depend on the size of processor memory and the number of timers.

FUNCTION BLOCK COUNTERS

Some P.C. systems, like Modicon's 84, 485 and 584, Allen-Bradley's PLC-3, and all Texas Instruments' P.C.'s, use a function block type format to program counters. You will notice that the function block for counters will be very similar in appearance to the timer function block. Even though their appearance is similar you will find the operation of counters a little different from the operation of timers.

In Fig. 7-2 you will notice that counters, just like the timers, have an enable terminal at the upper left side and a reset terminal on the lower left side. On the top right side there is a regular output and an inverted or NOT output at the bottom. The preset value, accumulative value, and register numbers are also similar. The

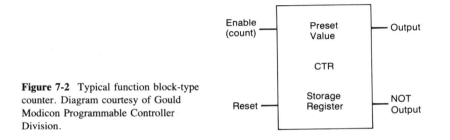

Figure 7-2 Typical function block-type counter. Diagram courtesy of Gould Modicon Programmable Controller Division.

major difference between the counters and timers is in the operation of the enable terminal.

ENABLE AND RESET

In Fig. 7-3 you can see contacts are added to the counter. As with the timer, these contacts are not part of the counter, but they are generally used to get the counter to perform a variety of functions. The enable contacts for this counter are 1004. You can select any input, as long as it can pass or change from open to closed. This contact transition from open to closed is what makes the counter count. You should remember that on the timer function block, the timer would begin to time the instant the enable contacts would close (rest must be energized or closed too). The timer actually timed how long the enable contacts stayed closed.

In the case of the counter, if the enable contacts change from open to closed, the counter adds 1 to its total, or accumulative count. Whether the enable contacts are closed for just an instant and then reopened, or closed for an hour and then reopened, the count would still be 1. This means that the counter only sees the electronic change at the enable terminal from open to closed, or 0 to 1. It does not matter how long this lasts as in the timer, rather it just counts how many times it has occurred.

For this reason some manufacturers refer to the enable contacts as *count*. This means that every time the count contacts go from open to closed, the counter will count or add 1 to its total.

The reset function on the counter block is very similar to the reset function on the timer. The reset terminal must be energized, or "high," for the counter to count. In most programs this is accomplished with a set of normally closed contacts, as in Fig. 7-3. Anytime contacts 1006 change from closed to open, or high to low, the counter resets the accumulative value to 0. In other words, anytime the reset contacts are opened, the counter returns to 0.

This also means that when you are finished counting and need to reset the counter, just open the reset contacts and send a low or 0 signal to deenergize the reset terminal. This will cause the counter accumulative value to return to 0 for the next events to be counted. These two terminals, enable and reset, allow the P.C. programmer to set up circuits to count a number of events, such as finished parts coming from a machine for one day, then to reset to begin counting the parts for the next day.

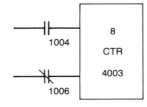

Figure 7-3 Typical counter with preset value and storage register number. Diagram courtesy of Gould Modicon Programmable Controller Division.

COUNTER PRESET VALUES
AND ACCUMULATIVE VALUES

The counter in Fig. 7-3 uses an 8 at the top to indicate a preset value, a CTR mneumatic in the middle to indicate this function block is a counter, and the number 4003 at the bottom to indicate the register where the accumulative count is being stored. These numbers and mneumatics become the most important part of the function block. The preset value can be any value that the P.C. manufacturer indicates fits in their processor's register. For instance, Modicon 84 and 484 will accommodate numbers up to three digits as preset values. This means that the largest number that fits into the registers is 999. The Allen-Bradley PLC-3 and Modicon 584 can take up to four-digit numbers. You must be sure to check the largest value your P.C. can handle to avoid program errors.

The preset value will indicate how many events that counter can count up to. This value is selected by the programmer and then entered into the function block through the programming panel or keyboard.

There are basically two ways to determine the preset value. One way is when you know that after a specified number of events, you want something to occur. For example, if the machine is stacking four boxes on a pallet, then placing a wrapper or banding strap around the pallet, you would use 4 as the preset value in the counter. When the counter counted to 4, indicating all four boxes were on the pallet, the counter preset would be met, and the output would signal the banding part of the machine to wrap the pallet.

The second way to program a counter is to count how many parts a machine is making during a day. In this case, since you don't know how many parts the machine will make but you need to keep an exact count, you would program the preset value higher than the machine's capability. This means that if the machine can produce about 500 parts per day, you would use a preset value of 999 and feel certain that the counter will keep an exact count. The only time your counter would lose count is if the machine produced more than 999 parts per day. If this is a possibility, then you will need to cascade the counter to a second counter. This technique will be discussed later. These two methods of programming preset values enable you to cover a variety of uses for counters.

The accumulative count is a very important part of the counter, but it does not show up anywhere in the counter function block. What does show up in the function block is the register location where the accumulative count is stored.

As in the timer function block, the counter will keep a running count with the accumulative value. When the accumulative value in the counter equals the preset value, the counter function block will energize the output terminal on its upper right side. Another important point to remember at this time is that when the accumulative value becomes equal to the preset value, the counter will stop counting. This means that once the accumulative value equals preset, then no more counts can be added to the accumulative value, even though the enable contacts continue to open and close. You will see how to acknowledge the overflow later.

Since it is important to know what the accumulative value is, you will need to examine the counter register. In the Modicon 484 and 584 you use the GET function just as you did for time registers. This means that you will have to move the cursor on the CRT to the GET area. This is in the lower right part of the screen for P-180 programming panels or the lower part of the screen for P190 programming panels. Once the cursor is in position, simply enter the counter's register number and use the GET key. You will then see the register number and the accumulative count show on the CRT. There are several example counters to program at the end of this chapter. You can increase their count by closing and opening the enable or count contacts and using the GET function to watch the accumulative count value change.

The number at the bottom of the counter function block indicates which register is being used to store the counter's accumulative values. The Modicon 84 provides 32 or 64 registers, the 484 provides 256, and the 584 provides 999 registers. Remember that these registers can be used to store timer data, counter data, or data math functions. Be sure to keep track of register assignments so that you don't try to program two functions, like a timer and counter, into the same register.

Counters in the Allen-Bradley PLC-3 and in Texas Instruments P.C.'s display the accumulative count or the current count. You will find that once you understand the relationship of the counter's preset value, accumulative value, and register number, you will be able to program the counter and use its outputs to control all types of machine operations.

COUNTER OUTPUTS USED FOR CASCADING

The counter function block has two outputs on its right side that operate identically as those on the timer function block. The output at the top right will be energized any time the accumulative value equals the preset value. This means the top output indicates that the counter has counted up to its maximum value and it is full. Some P.C. manufacturers refer to this output as the overflow output, since it shows that the counter is in an overflow condition. You will remember from earlier in this chapter that when the accumulative value equals the preset value, the counter stops counting and may become lost. Therefore, it is very important to recognize the overflow condition and to reset the counter when overflow occurs.

Figure 7-4 shows a sample program that acknowledges the overflow condition. The circuit operates like this. When enable or count contacts 1003 have opened and closed twenty times, the accumulative value of counter 4003 will equal its preset value of 20. At this point, the function block sends an output to coil 0006.

The normally open contacts of coil 0006 will close and add 1 to counter 4004, indicating an overflow has occurred in counter 4003. At the same instant, the normally closed contacts of coil 0006 will open and cause counter 4003 to reset. When counter 4003 resets, its accumulative value returns to 0 and it is ready for the next transition of enable or count contacts, input 1003. Since the accumulative

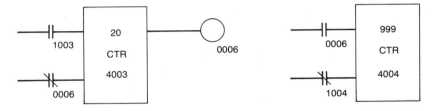

Figure 7-4 Using two counters to count values beyond 999. Diagram courtesy of Gould Modicon Programmable Controller Division.

value is now 0 and the preset value is 20, the output indicating overflow is now turned off, or deenergized. It will be ready to be energized again when the accumulative value counts up to the preset value, indicating the counter is in overflow again. You can see how this could count on indefinitely by adding another counter to catch the overflow of counter 4004. This type of programming is called *cascading*. Cascading can also be used to count up to very large values. For example, by setting the preset value at 999, you could count up into the billions using only four counters. You can see the possibilities, since you have many registers that can be programmed as counters. This shows several uses for the upper output that indicates an overflow condition has occurred. The lower output is also usable.

UNDERFLOW CONDITIONS

The lower output also operates like the output on the timer function. This output is energized anytime the accumulative value does not equal the preset value. For this reason, it is called the NOT output. Its operation is the opposite of the top output. Since the top output indicates an overflow condition when it is energized, the bottom output indicates the counter is not in an overflow condition. This is called an *underflow*.

The underflow output has uses similar to the overflow output. In fact, most programmers feel that the underflow output is redundant because identical functions can be achieved by sending the overflow output signal to a coil that has a set of normally closed contacts. You can prove this by changing the circuit in Fig. 7-4 to be reset with the normally open contacts controlled by a coil connected to the underflow output. Figure 7-5 shows this change. You should notice that coil 0007

Figure 7-5 Using the underflow (NOT) output to reset the counter. Diagram courtesy of Gould Modicon Programmable Controller Division.

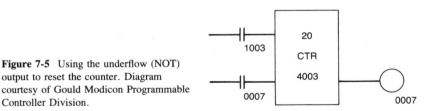

will be energized anytime the counter is not in an underflow condition. Since coil 0007 is energized the entire time the counter is counting, it keeps its normally open contacts 0007 closed to energize the reset terminal. As soon as accumulative value equals the preset value, the underflow output deenergizes the coil, and the reset contacts open, causing the counter to reset. You will notice that the same reset function has occurred, even though the overflow output was used in the first example and the underflow output was used in the second example. This type of redundancy sometimes confuses beginning programmers because it appears as though they will never learn all the possible ways to program. For now you can use either the underflow or overflow contacts and achieve the same results.

AUTOMATIC RESETTING COUNTERS

In some industrial processes you will need to program a counter that automatically resets. This may be required where you have a cycle that has several steps, like the example earlier in this chapter of stacking four cartons and then banding them with a wrapper. In this example, it is important for the machine to stack four boxes on a pallet, then signal the wrapping machine to band the four boxes together. At the end of this step, the counter should reset automatically and begin counting the next four boxes. When four boxes have been counted, the output signal is again sent to the wrapping machine to band these boxes.

Figure 7-6 shows a program that accomplishes the automatic reset. You will notice that the limit switch that counts the four boxes is connected to input 1003. This means that every time this limit switch closes and reopens, the counter in register 4003 adds 1 to the accumulative value. When the accumulative value equals the preset value of 4, an overflow condition is sensed by counter 4003, and an output signal is sent to output coil 0005. Output coil 0005 energizes and opens its normally closed contacts and automatically resets counter 4003. Output coil 0005 also powers a real output on the output module. Terminal 0005 on the output module is connected to the wrapper machine relay. This means that the output signal for 0005 powers a real output on the output module, turns on the wrapper machine, and opens a set of normally closed program contacts that reset counter 4003.

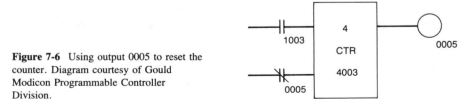

Figure 7-6 Using output 0005 to reset the counter. Diagram courtesy of Gould Modicon Programmable Controller Division.

From the example in Fig. 7-6 you can see that essentially any set of normally closed contacts that are connected to the reset terminal control the reset function of the counter. You should also remember that the reset contacts can be from a real-world input, such as a push button, or from a set of memory contacts that exist only in the processor or program as shown in Fig. 7-6. You will become more familiar with the operation of the counter by entering counter programs into the P.C. and testing their operation by operating the switch that controls the enable and reset functions. Sometimes though, you will have trouble entering a counter into the program or getting it to count and reset as you have planned. In this case, you should use a method to locate the exact problem, similar to the method explained in debugging a timer.

Once you fully understand the operations of the counter, you will be able to use any P.C.'s counters. Each counter you find will have similarities as far as preset and accumulated values, and types of outputs available.

For example, the Square D Company SY/MAX® P.C.'s counter is shown in Fig. 7-7. Notice that it has the abbreviation CTR to indicate this is a counter function. The up counter and down counter each have an enable terminal. If the up count terminal goes high, the accumulated count will be increased by 1. If the down counter terminal is energized, the accumulated count will be decreased by 1. The bottom terminal is called *clear*. This function resets the counter any time it is changed from high to low.

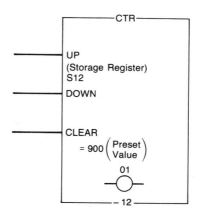

Figure 7-7 Typical Square D counter. Diagram courtesy of Square D Company.

The preset value is listed as the decode value. The accumulated value is stored in register S12. When the accumulated value equals the decoded value (preset value), the output listed at the bottom of the function block will be energized. By using closed contacts from this output, you can also show underflow conditions.

Hopefully you have understood counters well enough to understand how all counters operate. For this reason it is not necessary to show the counter program for each P.C. ever manufactured. You will soon learn to look for the similarities of new counters in the ones you already know. Be sure to complete the questions at the end of this chapter to test your knowledge.

QUESTIONS

1. Explain the operation of a P.C. counter, including the following:
   (a) Enable terminal
   (b) Reset terminal
   (c) Preset value
   (d) Accumulative value
   (e) Storage register
   (f) Output and NOT output
2. Name two things that determine the number of counters that can be put into a P.C. program.
3. Explain how you could make a set of counters count to values beyond 50,000.
4. What will happen when the accumulative value equals the preset value?
5. Explain how you could find out what the current or accumulative count is in a Modicon counter.
6. Explain how you could find out what the accumulative count is in an Allen-Bradley counter.
7. Write a program to show a counter counting to 4 and automatically resetting.
8. Write a program to show counters that will count to 35,000.
9. Write a program that will energize a lamp when the counter is full.
10. Write a program that uses a timer and several counters to keep track of minutes, hours, days, weeks, and months for a P.C. process.

chapter 8
Sequencers

Many industrial processes are sequential. For instance, some machines execute several steps in sequence during the operation. Most programmable controllers have some type of sequence function. Some P.C.'s have an actual sequence instruction, while others use the counter instruction with outputs to construct (program) a sequencer.

The sequence instruction allows the programmer to ensure that the machine operation stays in a stepped cycle. Some times this sequenced operation is required for safety, and other times it is needed for proper operations.

This chapter explains the operation of several types of sequencers. Several sample programs will be presented to demonstrate their operations. There will also be several programs for you to practice at the end of this chapter.

FUNCTION BLOCK SEQUENCER

An easy type of sequencer to understand is the function block format. Modicon 84, 484, and 584 use this type of function block. Figure 8-1 shows an example of a sequencer function block.

You should notice that the sequencer function actually uses a counter instruction. The difference is that the sequencer instructions are only located in eight counter registers numbered 4051–4058. Each of the eight sequencers have 32 independent steps available. (Some smaller units only have 16 steps.) Each step of a sequencer controls a set of contacts, which can be programmed normally open or normally closed.

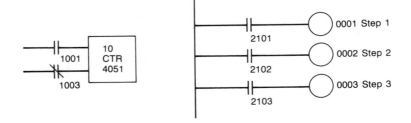

Figure 8-1 Typical Modicon sequencer function block with outputs. Diagram courtesy of Gould Modicon Programmable Controller Division.

BASIC SEQUENCER OPERATION

It may be easier to understand the operation of a sequencer in a sample program. Figure 8-2 shows a three-step sequencer program. This program would be used to sequence a press through its operation. On step one, the press moves its ram up. During step two, the ram moves down to open the mold. The third step energizes a cylinder to eject the parts.

The basic function block contains all the variables for the sequencer. The 3 used as the preset value indicates this sequencer will have three steps. (Any number of steps up to 32 could have been selected.) The CTR instruction indicates the function block will operate exactly like a counter. The register number 4051 indicates this counter is a sequencer. Remember, registers 4051 thru 4058 used with the CTR instruction become sequencers.

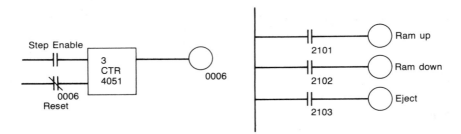

Figure 8-2 Typical 3-step sequencer program. Diagram courtesy of Gould Modicon Programmable Controller Division.

SEQUENCER CONTACTS

The contacts used to enable or advance the counter cause the sequencer to move through its steps in this case. These contacts can be controlled by a timer or limit switch, indicating the previous step is complete.

The output 0006 becomes energized when the enable contacts have been

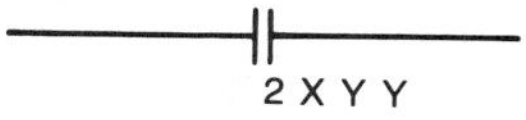

Figure 8-3 Modicon sequencer contact
numbering system. Diagram courtesy of
Gould Modicon Programmable Controller
Division.

```
 ──────────────┤├──────────────
               2 X Y Y
```

closed and opened three times. This part of the sequencer operates exactly like the counter. Output 0006 also powers a set of normally closed contacts used to reset the sequencer. The contacts marked 2101, 2102, 2103 are what actually make the sequencer different from the counter. The identification system for these numbers is very easy to understand. Figure 8-3 shows this system.

Starting on the left side of the number, the first digit, 2, indicates these contacts are special and are controlled by a sequencer. The next digit indicated by the letter X can be any number between 1 and 8. This number will correspond to the actual number of the sequencer that has control over these contacts. For instance the number 1 in the number 2102 and 2103 used in the sample program in Fig. 8-2 means that sequencer 1 controls these contacts. Sequencer 1 is stored in register 4051. If the number was 3, used such as 2032 or 2303 to identify these contacts, it would indicate that sequencer 3, stored in register 4053, would control these contacts. The last two digits in the sequencer's contact identifier number are represented by the letters YY. These two digits indicate which step of the sequence the contacts will be energized on. Any number between 01 and 32 can be used to identify the step number.

In the sample program in Fig. 8-2 numbers 01, 02, and 03 are used. This indicates that 2101 is energized during step 1, and 2102 is energized during step 2, and 2103 is energized during step 3. This program used the first three contacts in order. You are permitted to skip contacts if you would like, but the counter (sequencer) sequences through each step in order. Remember the actual number of steps in the sequence are determined by the preset value in the function block.

EXAMINING SEQUENCE STEPS

You may need to enter the program in Fig. 8-2 into your processor to get a better understanding of the sequencer. Use a push button as the enable contact. Use lamps as the outputs for each step of the sequencer. As you energize the enable push button, each lamp should light in sequence. Be sure the sequence number matches the contacts you are trying to control.

Since the sequence is programmed by using a counter (CTR) instruction, you can use the GET function to examine the sequencer's operation. Each time the enable terminal is energized, the accumulater's value increases by 1. When you used this register as a counter, the value displayed by the GET function indicated the count stored in the register. Now that you are using this register as a sequence, the accumulative number indicates the step the sequencer is currently on.

By using the GET function and watching the CRT to see which contacts are highlighted, you can determine which step the sequencer is on.

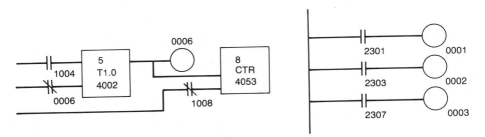

Figure 8-4 Variable timer incrementing a sequencer.

Figure 8-4 shows a sample program using variable times between steps. This program uses a 5-second timer in register 4002 to increment sequencer 3 through its eight steps. Since only steps one, three, and seven are used to control contacts, the time each contact is energized will be different.

The time between contact 1 and contact 3 being energized will be 10 seconds. The time between contact 3 and contact 7 would be twenty seconds, and the time between step seven and step one again would be 15 seconds. You might think that since there are only two steps between seven and one only 10 seconds would separate those two steps. But you should remember all Modicon sequencers start with step zero and return to step zero when they are reset. All outputs contacts controlled by the sequencer are deenergized during step zero. If you want the sequencer to skip step zero, you should use a one-shot or transition contact, which closes and opens at the speed of the processor's scan. This transition contact would be put in parallel with the sequencer's enable contact. It would be powered by the coil connected to the sequencer's output terminal. This is just another example of the possible uses for the sequencer function. With a little practice, you can learn to program sequencers to fit any application you may encounter.

USING THE ALLEN-BRADLEY SEQUENCER

Allen-Bradley P.C.'s, such as the 2/15, 2/20, and 2/30, use a function block to perform the sequencer operation. The sequencer function is used for the same type of applications indicated earlier in this chapter. The method of programming and operation will be quite different, though there will be some similarities to the Modicon sequencer.

The function block supports three sequencer instructions: sequencer output, sequencer input, and sequencer load. The function block transfers data from a data table to the output area. Each sequencer has up to 999 steps available. Each step can control up to 64 outputs. In comparison, the Modicon sequencer could only handle 32 steps, but had control over an unlimited number of outputs. Other differences and similarities will be covered in the sequencer's operation.

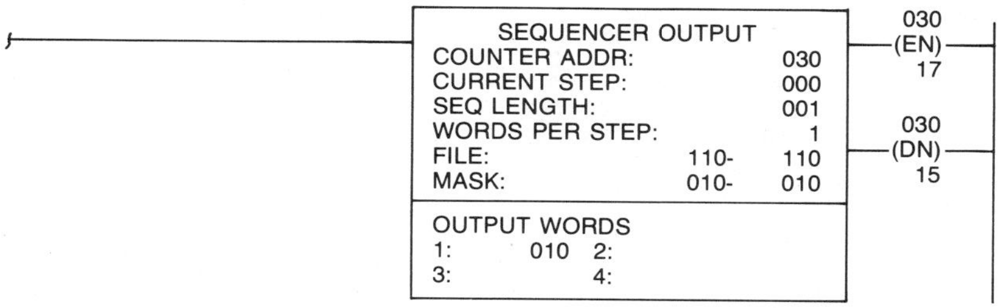

```
                          ┌──────────────────────────────────────┐      030
                          │        SEQUENCER OUTPUT                │    ─(EN)─
                          │   COUNTER ADDR:              030        │      17
                          │   CURRENT STEP:             000        │
                          │   SEQ LENGTH:               001        │
                          │   WORDS PER STEP:             1        │      030
                          │   FILE:              110-   110        │    ─(DN)─
                          │   MASK:              010-   010        │      15
                          ├──────────────────────────────────────┤
                          │   OUTPUT WORDS                         │
                          │   1:        010  2:                    │
                          │   3:             4:                    │
                          └──────────────────────────────────────┘
```

Numbers shown are default values. Numbers in shaded areas must be replaced by user-entered values. The number of default address digits initially displayed (3 or 4) will depend on the size of the data table.

COUNTER ADDRESS: address of the instruction in accumulated value area of data table.

CURRENT STEP: position in sequencer table (accumulated vlaue of counter)

SEQ LENGTH: Number of steps (preset value of counter)

WORDS PER STEP: width of sequencer table (number of columns)

FILE: starting address of sequencer table

MASK: starting address of mask file

OUTPUT WORDS: words controlled by the instruction.

Figure 8-5 Allen-Bradley sequencer output instruction. Diagram courtesy of Allen-Bradley Company.

SEQUENCER BLOCK FUNCTION

The sequencer is contained in a block function.

In Fig. 8-5 you can see that the block contains all pertinent information about the sequencer. The top line indicates that the function block is a sequencer. The next line indicates the address where the counter is stored. Remember the counter operation is incorporated into the sequencer's operation just like the Modicon sequencer's current step. This line will start at zero and display the current step number as the sequencer is incremented through its steps. The next line sequencer length indicates how many steps the sequence has. This is also the preset value of the counter. The next line indicates the number of output words the sequence controls. Remember each word is equal to 16 outputs. The outputs are numbered in octal (0–7 and 10–17). The next line indicates the starting address of the sequencer table file. The last line indicates the words that the instruction retrieves.

This may sound like so many words to you at this time. It will probably be easier to understand when you see an example program.

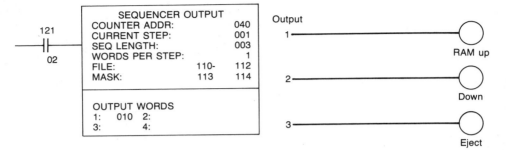

Figure 8-6 Allen-Bradley sequencer with outputs. Diagram courtesy of Allen-Bradley Company.

SAMPLE SEQUENCER PROGRAM

The sample program executes the industrial press operation presented earlier in this chapter. The machine had three sequential steps: ram up, ram down, and parts eject. Figure 8-6 shows the program as it would appear on the CRT after being entered into the processor.

You will notice the instruction, sequencer output, indicates any unprogrammed address could be used. The current step is displayed as 0, 1, 2, or 3. The 3 in sequencer length indicates this sequencer has three steps. Since only one output needs to be energized in each step, only one word is required for each step. Remember each word is actually 16 individual outputs. This means each step could have 15 outputs that are not being used. A mask could then be used to disable the 15 unused outputs in each word so memory won't be wasted. The file number 110–112 indicates the starting address of the sequencer data table in memory. You could use this address to examine the condition of larger, more complex sequencers. Since this sequencer has only three steps, each containing 1 output, you can observe the operation right from the function block.

The actual sequencer is enabled or advanced through the sequence by enable input 111/02. This means that every time contact 111/02 closes, the sequencer increments to the next step when it reaches the step. It then energizes the outputs indicated by the output word. In this example, the first time 111/02 closed, it would move the sequencer to step 1. Since step 1 controls only output 1, its output word would be 0000 0000 0000 0001. Only the bit identified by the 1 would be energized. All bits identified by 0s would not be energized.

When the enable contact opens, then closes for the second time, the sequencer will be incremented, or moved, to step 2. Since we want step 2 to energize output 2, its data word would be 0000,0000,0000,0010. Again the sequencer would energize all 1s in the output word and leave all 0s deenergized when it reached step 2.

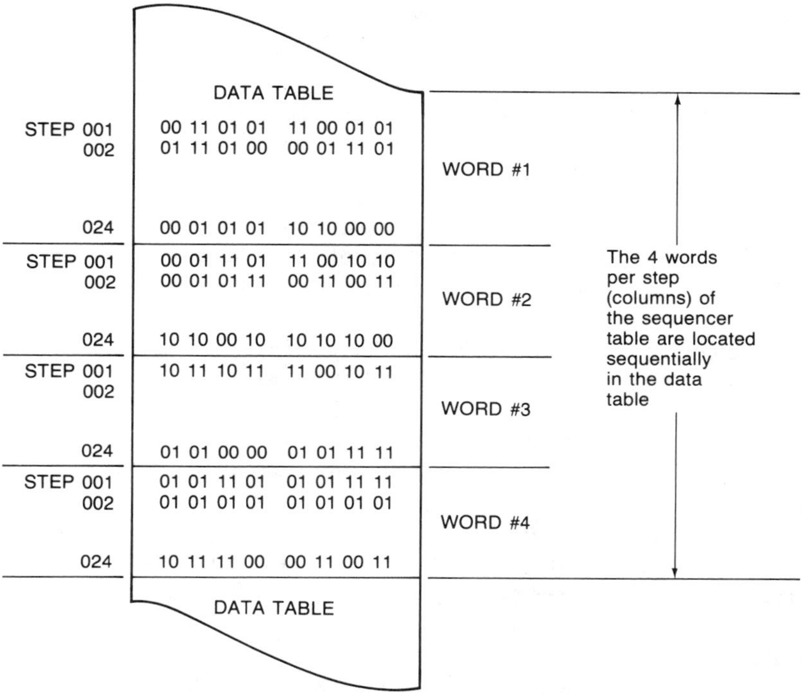

	DATA TABLE		
STEP 001	00 11 01 01	11 00 01 01	
002	01 11 01 00	00 01 11 01	WORD #1
024	00 01 01 01	10 10 00 00	
STEP 001	00 01 11 01	11 00 10 10	
002	00 01 01 11	00 11 00 11	WORD #2
024	10 10 00 10	10 10 10 00	
STEP 001	10 11 10 11	11 00 10 11	
002			WORD #3
024	01 01 00 00	01 01 11 11	
STEP 001	01 01 11 01	01 01 11 11	
002	01 01 01 01	01 01 01 01	WORD #4
024	10 11 11 00	00 11 00 11	
	DATA TABLE		

The 4 words per step (columns) of the sequencer table are located sequentially in the data table

Figure 8-7 Allen-Bradley sequencer data table. Diagram courtesy of Allen-Bradley Company.

The third time the enable switch closed it would move the sequencer to step 3. Now the current stop display in the function block would show a 3, indicating the sequencer is on step 3. Remember the number changes to show the current step.

Only output 3 would be energized during step 3 because its output word is 0000,0000,0000,0100. The sequencer output table can display all three steps together, showing which outputs would be energized during each step. Figure 8-7 shows the table for the sequencer. You can see that what is being called an *output word* is the group of sixteen 1s and 0s for each step. The processor knows to energize only the outputs that have a 1 indicated for that step. If the output has a 0, it will never be energized.

Once you understand the idea of the output word, the sequencer tends to be easy to understand and program. You should also keep in mind that each sequencer can have up to 999 steps and that during any step, the sequencer can energize any one of its 64 outputs. In fact, you could choose to energize all 64 outputs during one step. To do this you would need to place all 1s in the output word. Practice the Allen-Bradley type sequencers until you become proficient. Make some changes in the number of steps, the outputs you want energized during each step, and the

file addresses until you feel comfortable in their functions. Don't hesitate to refer
to your processor's technical manual for specific information.

CONSTRUCTING SEQUENCERS FROM COUNTERS

The Modicon and Allen-Bradley sequencers are most often found in programs.
Most other programmable controllers use one method or another that is similar to
these operations. There will be a few processors that do not have an actual sequencer
function or instruction listed on the keyboard. If the processor has a counter instruction and allows you access to the accumulative counter bits, you can construct
a sequencer.

The counters in most P.C.'s store the accumulated count value in a 16-bit
register. This register will use 12 of its bits to store the value in binary-coded
decimal (BCD). This is why most counters' preset value is limited to three decimal
values (up to 999). You should remember each decimal digit needs four binary
bits. Therefore, four binary bits for each of three digits will require a total of 12
bits for storage.

If the P.C.'s processor allows you to control an output on each of the 12 bits
you can start to see how these bits could be used to set up a sequencer. The sample
program in Fig. 8-8 shows how the output controlled by the first four BCD bits
could control up to sixteen outputs. Only the first ten outputs are controlled for this
example.

You can see each step controls one output coil. (Usually each output coil can
control an unlimited number of normally open and normally closed contacts.)

You should also know each output has four sets of contacts in series with
each other. These contacts are controlled by the four bits that make up the BCD
code. These are 8s, 4s, 2s, and 1s. You can see that the contacts will be programmed
as open contacts if the BCD bit is a 1 and they will be programmed as closed
contacts if the BCD is a 0 for that number.

When the counter moves to step one, the BCD value is 0001. This means
that only the combination of contacts for output 1 are all closed so that output 1 is
energized.

When the counter goes to step 2, the BCD value is 0010. You can see that
this time only the combination of contacts connected to output 2 are closed. The
combination of contacts connected to output 3 are all closed when the accumulated
count equals three. This sequence will continue through the count of 999. Remember, you would use the counter preset to select the number of steps that the sequencer
would require.

By now you are familiar with your processor's instructions and the way they
operate. You can see that the most important part of the P.C. system is the person
that will program, troubleshoot, and maintain the system. Even though the P.C. is
totally automated, it still requires a human to program it.

Check out the problems and questions at the end of this chapter to ensure
that you have comprehended the material in this chapter.

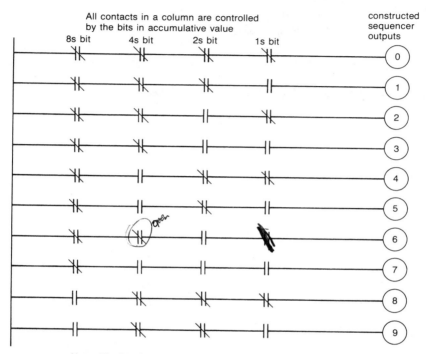

Figure 8-8 Constructing a sequencer from four BCD bits of the counter accumulative value.

QUESTIONS

1. Name the counter registers in the Modicon 484 that have sequence operation available.
2. Explain how a three-step sequencer operates.
3. If you wanted a three-step sequencer, what value would you use as the Preset value?
4. Explain what each digit of the number in the Modicon contacts ─┤├─ 2101 means.
5. Explain how the Allen-Bradley timer operates.
6. How can you tell what step a Modocon sequencer is on?

7. How can you tell what step an Allen-Bradley sequencer is on?

8. Explain what happens in a ten-step sequencer if the reset line is opened and closed in step 4.

9. Explain how you can get 5-second delay between the first three steps of an eight-step Modicon sequencer.

10. How many contacts in Modicon program can have the number 2101?

11. Write a program that shows a four-step sequence. (Use four different lamps for outputs for the four steps.)

12. Change the program in problem 11 so that there is not a step where all four lamps are off.

13. Write a program for an eight-step sequence that has a 10-second time delay between steps 4–5 and 20 seconds between steps 6–7.

chapter 9
Math Functions

Mathematic functions like addition, subtraction, multiplication, and division allow the programmable controller to handle numbers for data gathering and comparisons. These functions give the P.C. power similar to normal computers. In this case, power refers to the ability to calculate and make decisions similar to a human being, based on the data in memory. This chapter will explain how to program and use the four basic mathematic functions on P.C.'s in general. Examples will be presented that show these functions in programs for Modicon and Allen-Bradley P.C.'s. Since other P.C.'s are similar, sample programs are presented without explanations. You will begin to see that all P.C.'s handle math functions in a similar manner. Only small format changes make them different. You will also see the fundamental uses for math functions and be capable of applying these in your own applications. As in previous chapters, you should practice the programs that are presented, if you have a P.C. available. You may need some minor program modification if you are using a P.C. that is not discussed here. It should also be pointed out that some small programmable controllers do not provide math functions. In general all manufacturers provide math functions on their larger P.C.'s.

ADDITION FUNCTION

The simplest math function to understand is the addition function. The addition function allows the P.C. system to keep track of events, like total part count. If you remember, in the chapter discussing counters, we programmed a counter to count the parts that a machine produced each day. We can now use the addition

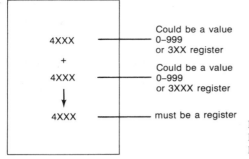

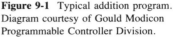

Figure 9-1 Typical addition program.
Diagram courtesy of Gould Modicon
Programmable Controller Division.

function to expand that program to calculate the total parts produced for a week. The addition function can also be used if you have more than one machine making the same part. For instance, if you have three machines that are producing the same part, you would need to add the count of each machine to calculate the total parts produced. These two examples demonstrate several uses for the addition function. Now we will see how the addition function is programmed and executed.

A simple type of addition function to understand is the function block type. The function block format is used by Modicon on its Micro 84, 484, 584, and 884, and by Allen-Bradley on its PLC-3, and by Texas Instruments 520 and 530 processors, as well as many others. Figure 9-1 shows the function block for a Modicon 84 and 484. You will notice that math function blocks use three rungs of program. The value in the top of the block is mathematically added to the value in the middle of the block. Register numbers can be used instead of values in the top and middle of the block. This means that if a 4XXX number is used, the processor checks to see what value is stored in register 4XXX and uses that value in the addition process.

When the top value is added to the value in the middle line of the function block, the answer is stored in the register listed on the bottom line. This means that the format for the addition function block allows any value up to 999 or any register (4XXX or 3XXX) number to appear on the top and middle lines. The bottom line must be a 4XXX register number.

This function block has only one input terminal on the top left side of the function block. The processor executes the addition of the top and middle lines in the function block when the input terminal is energized or "high." This may also be called the enable terminal. The actual addition is executed every time the processor scans the program. The answer remains the same during each scan unless one of the values changes. The addition function block has 1 output on the right side. The top output is energized when the answer to the problem is larger than 999. The function block output allows the programmer to program the P.C. to take action (energize outputs) when register overflow conditions exist. It is also possible to program math functions to perform calculations, like addition and then simply store the answer in a register for future use. This option gives the programmer many possible uses for the addition function.

Figure 9-2 shows a sample program where a carton containing four parts passes a counter before being placed on a pallet. Register 4004 is used to count the total number of parts that are being shipped out on the pallets. This means that each time a carton passes the counter, a 4 is added to the total in the register 4004.

A limit switch is used to count the cartons as they pass. It is connected to input 1003. The switch closes and opens each time a carton passes and energizes contacts 1003. The transitional contact limits the processor to completing the calculation only once each time the switch is closed. Each time 1003 is energized, a 4 is added to value register 4004. (Remember, any 4XXX register can be used as long as it is not assigned use as a counter, timer, or other function in this program.) When the program starts, register 4004 has a value of 0, indicating no parts have been counted. When the first carton passes the limit switch, a 4 is added to the contents of register 4004. In this case, 4 + 0 = 4. Now register 4004 has stored a value of 4. When the second carton passes limit switch 1003, the addition takes place again, adding 4 to the contents of register 4004. This time the function block adds 4 to the stored value of 4, and the answer 8 is placed in register 4004. Each time, the processor adds 4 to the previous value in register 4004. The answer is stored right back in register 4004, which causes register 4004 to accumulate the total. When you try this program on your processor, be sure to use the GET function on register 4004 so you can see it update the total each time 1003 is energized. If you think more than 999 parts will be counted, you should use an output to indicate that an overflow has occurred. This overflow is usually "dumped," or programmed into a counter. Another way to keep track of numbers larger than 999 is to use double precision math registers. This technique is explained later in this chapter.

Another sample program using the addition function is shown in Fig. 9-3. In this example, the program is adding the value in register 4002 to the value in register 4006. The answer is stored in register 4008. Register 4002 is storing the total count for counter number 1, indicating the number of parts that machine number 1 has produced. Register 4006 is storing the total count for counter number 2, indicating the number of parts that machine number 2 has produced. The addition function is adding the current count from register number 4002 to register 4006 and storing

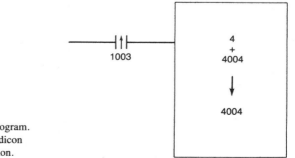

Figure 9-2 Sample addition program. Diagram courtesy of Gould Modicon Programmable Controller Division.

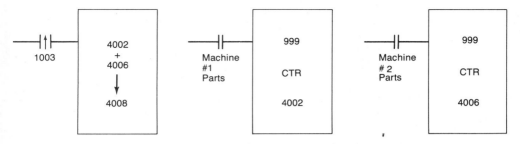

Figure 9-3 Math program used to add counter 4002 to counter 4006. Diagram courtesy of Gould Modicon Programmable Controller Division.

the result in register 4008. For this example, register 4008 indicates the total number of parts that both machines have made.

You will notice that the addition function can be programmed to constantly update the total by powering the enable terminal constantly. If scan time was critical, you could use a push-button panel switch connected to input 1003 to enable the addition function block in Fig. 9-3. This way the processor would only compute the total when someone pressed the push button, thus saving scan time. Another option would be to use a timer to energize the function block periodically updating the total. You should also notice that any change in either counter will cause a change in the total.

These sample programs show you several variations in the program and use of the addition function. With practice, you will be able to tailor these programs to your applications.

SUBTRACTION FUNCTION

The subtraction function is very similar to the addition function. The major differences are that the subtraction function is more versatile and usually controls one or more outputs. Some programmable controllers use a set of comparison functions instead of subtraction. These functions, less than, equal to, and greater than, create the same operating conditions as the subtraction function. This portion of chapter 9 demonstrates the operation of the subtraction function and the comparison instruction.

The easiest type of subtraction function to understand is the function block. Modicon 84, 484 and 584 use this type of programming format, as well as many other P.C.'s. Figure 9-4 shows a simple example of the subtraction function. The subtraction function uses three program rungs, just like the addition function block. The top and middle elements in the function block must be a value of 0–999 or a register numbered 4XXX. If a register number is used, then the value stored in that

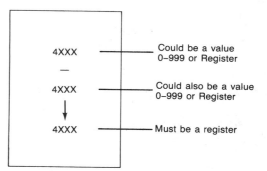

Figure 9-4 Typical math function block used for subtraction. Diagram courtesy of Gould Modicon Programmable Controller Division.

register will be used for all calculations. The bottom element can only be a 4XXX register number.

The operation of the subtraction function takes the value of the middle element (or value stored in the register) and subtracts it from the value in the top element (or value stored in the register). The answer is then stored in the register indicated by the number of the bottom element. Sometimes the answer is called the *difference*.

The subtraction function block has only one input located at the upper left terminal. Its operation is similar to the addition function in that the actual subtraction calculation only occurs when the terminal is energized. When the input is energized, the subtraction calculation is updated each time the processor scans the program.

The subtraction function controls 3 outputs that are located on the right side of the function block. The top output is energized when the upper value in the function block is greater than the middle value. If 4XXX register numbers are used, then the processor compares the value stored in those registers. The middle output

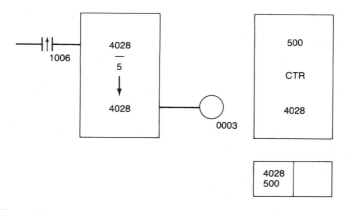

Figure 9-5 Subtraction program with results shown in reference area. Diagram courtesy of Gould Modicon Programmable Controller Division.

is energized when the upper value and middle value are equal to each other. The lower output is energized when the middle value is less than the upper value. You can now see that the function block is able to subtract, or show the difference, between two values. The answer is stored in the register indicated by the lower element. It can also be used to make comparisons of greater than, less than, or equal to, and to energize appropriate outputs to indicate these conditions.

It might be easier to understand these functions by programming a sample program. Figure 9-5 shows a sample subtraction program. In this example, a machine is molding five parts at a time. You want the machine to make 500 parts, so 500 is placed into register 4028. It is important to understand that Modicon P.C.'s do not have plain, ordinary registers. You can make something similar to registers by using the counter (CTR instruction). Since you do not have enable contacts, the register does not actually count down, instead it simply stores a value. Another point should also be made at this time. You should notice that the CTR register and the subtraction function block are both using the same register, 4028. This practice works fine for this program because the CTR function does not operate and change the value. Only changes in the subtraction function alter the value stored in 4028. You must use the GET function to place the value 500 into register 4028.

After programming the example in Figure 9-5, use the keyboard to move the cursor to the GET area at the bottom reference area of the CRT. Retrieve register 4028 in the GET area, and place the value into the register by pressing the 500 key and the ENTER key. (You can use a smaller value if you wish.) Now register 4028 has stored the value 500.

The program operates similar to the addition program. Limit switch 1006 energizes each time five parts are made. When switch 1006 closes, 5 will be subtracted from the value stored in register 4028. This occurs because the subtraction function subtracts the middle value from the top value and puts the answer in the register indicated at the bottom of the block. In this case, the answer becomes the value stored in register 4028. This means that during the first cycle, 5 is subtracted from 500. The answer 495 is now placed in register 4028. When the subtraction takes place the second time, 5 is subtracted from 495. The answer 490 is now placed in register 4028. Each time the limit switch closes, the number in 4028 is reduced by 5. This operation continues until the value in 4028 is less than 5. At that time, the subtraction function block energizes the bottom output terminal. This terminal powers a coil that opens a set of normally closed contacts and shuts the machine off.

You may wonder why this program would be used if the addition function could perform the same task. Most programmers like this method because the number of parts to be produced can be entered into register 4028, and the machine is controlled by the subtraction function. In a later chapter, you will see that a thumbwheel can also be used to enter the value directly into register 4028 without the use of the programming panel. The thumbwheel can be mounted for easy access and allows on-the-spot changes in production quantities.

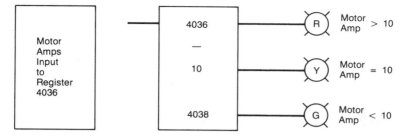

Figure 9-6 Subtraction function used for comparison. Diagram courtesy of Gould Modicon Programmable Controller Division.

USING THE SUBTRACTION FUNCTION FOR COMPARISON

The real strength of the subtraction function is its ability to compare two values. The function block is constantly comparing the top element to the middle element and using the outputs to indicate if the top element is greater than, equal to, or less than the middle value.

Figure 9-6 shows the subtraction function used as a monitor to keep track of motor current. The current for a motor is continually entered into register 4036 (an explanation of this operation will be presented in a later chapter). The value 10 in the subtraction register is used as a set point indicating that you want the motor to run using less than 10 amps. The subtraction function is enabled continually with the shorting bar, ensuring that the comparison occurs each time the processor scans the program, which occurs many times each second. During each scan, the subtraction function is executed, and the motor current is compared to the set point, 10 amps. If the motor current in register 4036 is less than 10 amps, then the bottom output energizes a green lamp, indicating motor current is in a safe range. If the motor current increases to exactly 10 amps, the subtraction function indicates that the top and middle value are equal and energizes the middle output. The output energizes a yellow lamp, indicating the motor current is nearing an unsafe range. If the motor current becomes greater than 10 amps, the top output energizes a red lamp, indicating motor current is now too high.

You can simulate this program operation without a motor current by altering it as shown in Fig. 9-7. Make 4036 into a counter register with a preset of 15. Use a push-button 1003 to increase the count or value in register 4036. This simulates an increase in motor current. You will notice that as long as the value in the counter 4036 is less than 10, the upper output is energized, indicating motor current is safe. When the value in 4036 is exactly 10, the middle output is energized, indicating the current is equal to the set point and the middle output on the subtraction function block will be energized indicating motor current is safe. Use the reset button to return the motor current (counter) to 0.

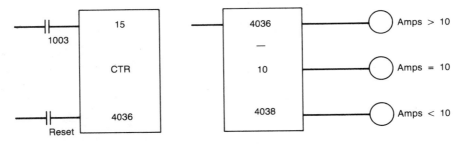

Figure 9-7 Subtraction function used for comparing a register value to a constant. Diagram courtesy of Gould Modicon Programmable Controller Division.

You can see that the subtraction function is very powerful. You can use it to compare temperatures, liquid levels, bin weights, pressures, and other production and process variables to their set points. An indicator lamp is used when conditions are satisfactory; if conditions were unsafe, the outputs energize values and solenoid or other controls.

FUNCTION BLOCK MULTIPLICATION

From Fig. 9-8 you can see the function block used for multiplication has three elements, just like the other math functions. The top and middle elements can be either a three-digit number up to 999 or a register number 4XXX. The bottom element is a 4XXX register number where the answer to the multiplication problem is stored. It is important to note at this time that the multiplication function actually designates two registers for the answer. This means that the register number listed on the bottom line of the multiplication function block and the very next register are used to store the answer. (Be sure both registers are not being used by any

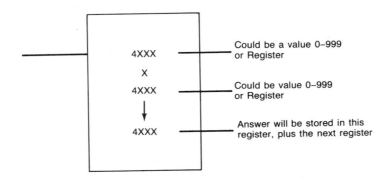

Figure 9-8 Typical multiplication function block. Diagram courtesy of Gould Modicon Programmable Controller Division.

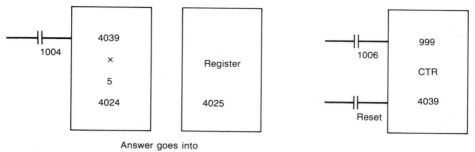

Answer goes into
registers 4024–4025; register 4025 will not be displayed on CRT.

Figure 9-9 Double precision registers used for multiplication. Diagram courtesy of Gould Modicon Programmable Controller Division.

other function.) Each register will only store up to three digits, so the largest answer the multiplication function can handle is 999 999.

Figure 9-9 shows a typical multiplication program using double precision registers. This example is another way to count the parts coming from the machine used in the subtraction sample program. Since the machine produces five parts each time a mold is ejected, you can simply count the number of times the machine has ejected parts, using limit switch 1006 to enable the counter in register 4039. When you need to know the total parts produced, you can close switch 1004 to energize the multiplication function. The multiplication function multiplies the total number of times the machine has ejected parts, which is stored in 4039, by five. The answer is placed in registers 4024 and 4025.

For example, if counter 4039 has a value of 785, the multiplication function multiplies 785 five times. The answer to this calculation is 3925. Since this answer is too large to fit into one register, it is broken into groups of three digits and placed in registers 4024 and 4025, as shown in Fig. 9-10. You should notice that the answer 3925 has four digits. Therefore when it is broken into groups of three digits,

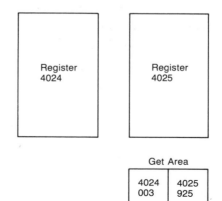

Figure 9-10 Double precision multiplication registers with answer shown in reference area of CRT. Diagram courtesy of Gould Modicon Programmable Controller Division.

two zeros must be added to the left side of the number 3925 making it 003925. Notice that the zeros do not change the value, rather they indicate that space has been filled. You should also notice that the three most significant digits are stored in the register listed at the bottom of the multiplication function block, while the three least significant digits are stored in the next register. For example, since the answer was 003925, register 4024 stores 003 and register 4025 stores 925.

At first this may seem a little difficult, but with practice you will soon begin to understand the process. The best way to practice the use of double precision registers used in the multiplication function is to use the program example listed in Fig. 9-9. You should use the GET function to retrieve the registers listed in the bottom line of your function block plus the very next register. As a rule, it is best to place the lower-numbered register on the left side of the higher-numbered register. This way the digits in the answer are side by side as you would actually read them. Next use the GET and ENTER keys to place values into the counter register. Then close the switch to enable the multiplication function, and check the registers in the GET area of the CRT to see the answer. Be sure to do the multiplication problem on paper or with a calculator and divide the digits in the answer into groups of three. This way you should see the same numbers that are in the two answer registers.

You may need one more example to fully understand this process. For instance, if you used the values 385 times 45 in your calculation, and register number 4048 was listed on the bottom element of the function block as the storage register, you should find the value 017 in register 4048, and 325 in register 4049. The values 017 and 325 indicate the answer for the calculation 385 $\times$ 45 is 17325.

USING MULTIPLICATION FOR SCALING

You have probably begun to notice that you can use the multiply function either of two ways; the contents of one register can be multiplied by the contents of another register, or you can multiply the contents of a register by a set value. Many times values brought into the P.C. registers are not full scale. For instance when a load cell is used to indicate the weight of a bin, it generally sends a signal that is accurate but not full scale. When this occurs, the multiplication function can be used to bring the value to full scale. In this case, the load cell sends a signal indicating a value of 400 when the bin actually weighs 1200 pounds. Since the signal is scaled by 1/3, you could get the programmable controller to indicate actual bin weight by multiplying the signal value by 3. The number used as a multiplying factor is sometimes called a *fudge factor*, indicating it helps change the signal back to an accurate full-scale value.

You should begin to see many uses for the multiplication functions. You can see that it is especially useful to scale values. As you become more familiar with its operation, more uses will become apparent.

OUTPUTS ON THE MULTIPLICATION FUNCTION BLOCK

The multiplication function block has only one output, and it is located on its top right side. It is energized any time the value in the storage register, listed as the bottom element of the function block, is larger than 999. In other words it is energized any time a register overflow has occurred. If the answer is smaller than 999, the output is not energized. It is also important to note that the output can only be energized when the multiplication function block enable terminal is energized. This output terminal can be used to indicate an overflow or be connected to a coil for control purposes.

DIVISION FUNCTION

The division function is the fourth math function supported by most programmable controllers. Its operation and programming are very similar to the multiplication function. It also is a little more complex than the subtraction and addition function because it uses double precision registers or two data bytes for the calculation. An easy type of division program to understand is the function block format used by Modicon P.C.'s. Figure 9-11 shows an example of a function block-type program. You will notice the function block uses three elements or program lines, just like the other math function blocks. The top element and bottom element in the function block must be 4XXX register numbers. The middle element can be a value 0–999 or 4XXX register. The format indicates that the top element is divided by the middle element and that the answer is stored in the register indicated in the bottom element. Just as in the multiplication function, numbers can get rather large. For this reason, double precision registers are also used in the division function block. In this case, the register listed in the top element of the function block uses the double precision registers. This means that the division function uses the register listed in the top element, plus the very next register too.

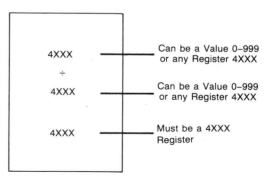

Figure 9-11 Typical division function block. Diagram courtesy of Gould Modicon Programmable Controller Division.

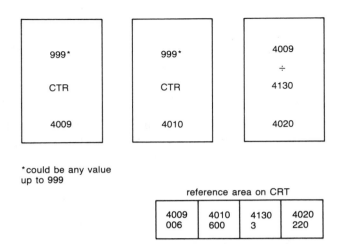

*could be any value
up to 999

reference area on CRT

4009	4010	4130	4020
006	600	3	220

Figure 9-12 Example division program with results shown in the reference area of the CRT. Diagram courtesy of Gould Modicon Programmable Controller Division.

In this example, the value in double precision registers 4009 and 4010 is divided by the value in register 4130. The answer is put in register 4020. From the GET function shown on the bottom of Figure 9-12, you can see that the division calculation will be $006600 \div 3 = 220$. In this case register 4009 has the value 006, and register 4010 has 600. The actual value in these double precision registers is 006600. Each time the processor scans the program, the calculation is completed. The output is energized any time the answer is too large for the division register.

ALLEN-BRADLEY FUNCTIONS

The Allen-Bradley addition problem is shown in Fig. 9-13. In this problem contact 111/11 is used to enable the math function. The processor uses the GET function to complete the addition. In the Allen-Bradley P.C., the GET function is used because the processor must go to the register listed at the top of the contact and get the value. This value is then listed at the bottom of the GET function. The math function at the output (in this case, +) takes the values in data word 030 and 031 and adds them together. The answer is placed in data word 032.

Figure 9-13 Allen-Bradley-type addition function. Diagram courtesy of Allen-Bradley Company.

You can see from this example that the value 520 in data word 030 and the value 514 in data word 031 are added together and placed in data word 032. The answer in this example is 1034.

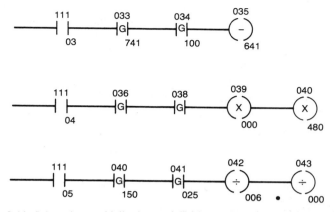

Figure 9-14 Subtraction, multiplication, and division program for an Allen-Bradley system. Diagram courtesy of Allen-Bradley Company.

Other examples of subtraction, multiplication, and division are shown in Fig. 9-14.

By now you can see that math functions are fairly easy to understand. Basically, values from one register are added to values in another register, or some other math function is completed.

COMPARISONS

Some P.C.'s use a simple comparison function block. Figure 9-15 shows examples of these three comparisons. The blocks basically produce an output any time the comparison is true.

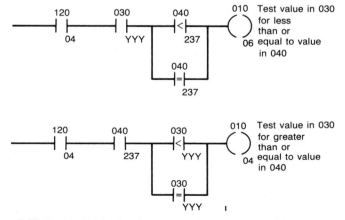

Figure 9-15 Testing 030 for less than, greater than, or equal to. Diagram courtesy of Allen-Bradley Company.

If the comparison function is not true, the output is not energized. By grouping the greater than, less than, and equal to symbols, you can test for nearly any possible condition.

Some processors can perform advanced math functions like squaring and square roots. In these systems, the advanced math functions work similarly to other math functions. All you need to know is how the format works.

You may need to try the problems and programs at the end of this chapter. You will soon see that most P.C.'s can easily integrate math and comparison functions with normal I/O control.

QUESTIONS

1. Show the function block program for addition function.
2. Explain how the addition function works.
3. Show the function block program for a subtraction function.
4. Explain how the subtraction function operates.
5. Show the function block program for a multiplication function.
6. Explain how the multiplication function operates.
7. Show the function block program for the division function.
8. Explain how the division function operates.
9. What are double precision registers?
10. Explain how Allen-Bradley uses the GET and PUT functions for math operations.
11. Write a program using the function block type program for
    a. Addition
    b. Subtraction
    c. Multiplication
    d. Division
12. Write an Allen-Bradley program for two math functions.
13. How can you see if the math function is operating properly in the program?
14. Explain how the processor handles an addition function block that has a register number on the top line.
15. Explain the use of the subtraction function block as a comparison function.
16. Write a program using a subtraction function block that compares two values for greater than, equal to, and less than.
17. What is the output on the addition function block generally used for?

chapter 10
I/O Devices

In chapter 5 you saw contacts and coils represented in programs by symbols. The most important concept to understand when learning about a programmable controller is the relationship between these contact and coil symbols in the program and the actual switches and motors found on the industrial machinery.

As you learned in earlier chapters, the contacts and coils in the P.C. program can be connected to the "real world" to receive inputs from switches and to send outputs to coils and motor starters, or they can merely exist in the P.C.'s memory, where they are simply a programming convenience. Since both real-world and memory symbols look the same on the CRT when you view the program, it is sometimes confusing for someone trying to learn P.C.'s. This chapter uses a variety of diagrams, explanations, and examples to help you understand their differences. Its basic intent is to provide you with an understanding of inputs and outputs. It demonstrates their differences with memory coils and contacts. This discussion begins with inputs and continues onto outputs.

BASIC INPUT TERMS

The word *input* is used in a variety of ways to describe some aspect of electricity or electronics. Since the word usually has several meanings, it is very helpful to discuss them so you can understand them more easily. First of all, any switch that is used to send electrical signals can be called an input. It more accurately should be called an *electrical input device*. The electrical signal that the switch sends could also be called an input. It, too, would be easier to understand if the term *input signal* were used. The electronic circuit on the programmable controller that receives the input signal is also called an input. Its more precise name is *input module*.

In the explanations in this chapter, the term *input device* is used to describe push-button, selector, limit, and all other types of switches. The term *input signal* refers to the electrical impulse that flows between the switch and module, and between the module and processor. The term *input module* describes the electronic circuit that receives the input signal at the processor. It is called a module because it refers to several independent electronic circuits housed in one plastic case. It is easier to understand how these three parts, the input device, input signal, and input module, all operate together if you know how the input module works.

BASIC INPUT MODULE

Figure 10-1 shows the electrical diagram for a typical input module. It would be too difficult to describe all the circuits used by all programmable controller man-ufacturers. This circuit was selected because it is very typical of all input module circuits and because you can actually construct this circuit using the values indicated and test the concept firsthand.

The basic function of this circuit is to send a low voltage signal that is usually at the 5 volt TTL level to the processor on terminals A and B when a high voltage signal is received at terminals 1 and 2.

From the diagram you can see that the high voltage always comes from a source outside the input module. In most industrial applications, this source is a transformer or power supply that provides 24, 110, or 220 volts. The input device (switch) is connected in series with the neutral terminal of the transformer, while the "hot" side is connected directly to terminal 1 of the module. NOTE: The hot and neutral wires may be reversed by some P.C. manufacturers. Check your sys-tem's specifications to be sure.

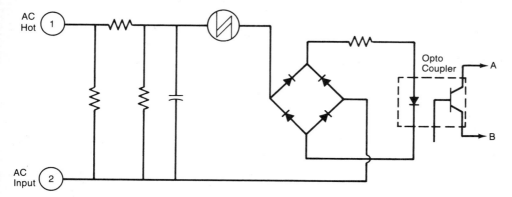

Figure 10-1 Wiring diagram for typical input module. Diagram courtesy of Gould Modicon Programmable Controller Division.

It should be quite clear by now that the high voltage power source and the input device are important to the operation of the input module. But remember, they are not part of the input module. In fact they may be located quite some distance from the input module. A typical input module is shown in Fig. 10-2. From the picture you can see the electronic circuit is housed inside a plastic case.

The circuit begins to operate when the switch closes and completes the circuit between the power supply's neutral terminal and terminal 2 on the input module. The first thing that happens when terminal 2 is connected to the neutral terminal is that the neon indicator lamp is energized. This lamp is placed in the input module so that the glow from the lamp is highly visible outside the module. You can be certain that the input module is receiving a neutral signal when the neon indicator is glowing.

Since the neon indicator lamp is connected in parallel, the neutral terminal also completes the circuit through a voltage-dropping resistor with the four diode full-wave bridge rectifier. In this application, the diode bridge rectifier is changing the AC voltage signal to DC, or direct current signal. The DC output of the bridge rectifier is carried through a 510 ohm resistor and a light-emitting diode (LED). The light-emitting diode's output is infrared light. The light-emitting diode is part

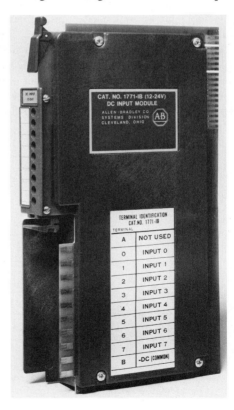

Figure 10-2 Typical input module. Picture courtesy of Allen-Bradley Company.

of an electronic component called an *opto coupler*. The opto coupler contains two parts: a light-emitting diode encapsulated in plastic with a phototransistor. The light-emitting diode is placed inside the plastic so its light shines directly onto the phototransistor. The opto coupler is fully encapsulated in plastic so that outside light will not interfere with its operation. When the DC signal passes through the LED, it becomes energized and shines light on the phototransistor. The photo-transistor acts as a switch. Regulated DC is present at terminal A of the photo-transistor. When light from the LED shines on the phototransistor, it switches on and passes current through to terminal B. This current becomes the input signal to the processor.

Essentially this means that any time the input device (switch) completes the circuit, the LED will be powered and the phototransistor will conduct and allow the low voltage signal to pass into the processor. At the same time, the neon indicator will become illuminated indicating an input signal is present.

Figure 10-3 shows a simplified diagram of the input switches and input module. The action of transmitting the signal by using the light of the LED effectively isolates the high voltage part of the circuit from the low voltage part. For this reason, the opto coupler is sometimes referred to as an *opto isolater*.

This simplified diagram is generally the type of diagram most manufacturers use. It does not show the electronic circuit containing the opto coupler. Instead, only the terminals for the input voltage, input switch, and module terminals are shown. The manufacturers assume you understand what is going on inside the module's circuit, therefore, you only need to know where to connect the input's high voltage power supply and the input switch. From Fig. 10-3 you can also see some manufacturers show the exact number of input circuits in the module. These diagrams also become the actual wiring diagram the technician uses to install the

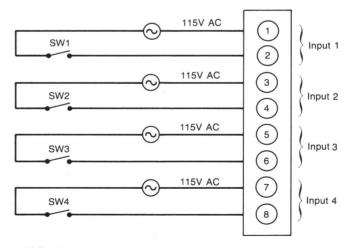

Figure 10-3 Field wiring diagram for 115-volt input module. Diagram courtesy of Gould Modicon Programmable Controller Division.

input module. It might be good to remember that input modules generally have a printed circuit board edge connector to make the connection into the input/output module rack. This means the connection to the rack that sends the signal to the processor requires no wiring. Simply push the module into the rack until the edge connector is seated. Since the input module's connection to the rack is rather foolproof, the manufacturers generally only include the diagram for the wiring that the technician must complete on the switch side.

OUTPUT MODULES

Output modules look very similar to input modules in appearance. Figure 10-4 shows a picture of an output module. This is because both the input and output modules can be connected in the same rack. As with the input side of the programmable controller, the output side contains several distinct and separate parts that are all usually referred to as the *output*. Since this is somewhat confusing to someone trying to learn how a P.C. system operates, more definitive names will

Figure 10-4 Typical output module. Picture courtesy of Allen-Bradley Company.

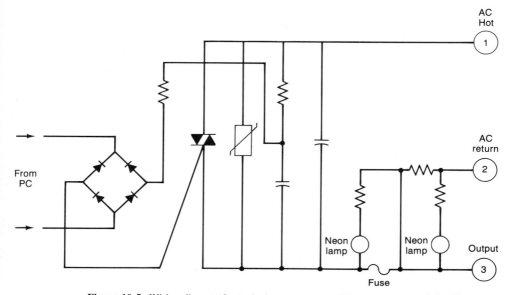

Figure 10-5 Wiring diagram for typical output module. Diagram courtesy of Gould Modicon Programmable Controller Division.

be used in this chapter. The signal coming out of the processor will be called an *output signal*. The electronic circuit in the module will be called the *output module*, and the electrical load that receives power from the module will be called *output load*.

The source of power to energize the output load is not contained inside the output module. From the diagram in Fig. 10-5 you can see it is a separate part, usually a transformer. When the processor sends a signal to the output module, it electronically closes the circuit and passes current to the output load. Figure 10-5 shows a typical electronic circuit for an output module.

From Fig. 10-5 you can see that the "hot" and neutral terminals of the transformer are connected to terminals 1 and 2. One side of the electrical load is connected directly to the neutral terminal of the transformer. Power from the hot side of the transformer is switched on and off by the *triac*. The triac is an electronic switch that turns on any time a signal is received at its gate.

The signal to turn on the triac is sent from the P. C.'s processor. The processor sends a signal through an opto coupler to isolate the computer from the outside electrical world. Remember, opto couplers or opto isolators are also used in output modules to protect the fragile processor from large current and voltage surges that are present in all industrial electrical systems.

The signal that the processor sends out is called an *output signal*. As this signal passes through the opto coupler, it energizes the phototransistor so it completes a path through the DC portion of the four-diode bridge rectifier. Once the DC side of the bridge rectifier allows current to flow, a small current also flows in the AC side of the bridge. This current is used to provide gate circuit to the triac. In this circuit, the diode bridge rectifier is not used to rectify or change AC voltage to DC voltage. Instead, the bridge is used like a switch. When current is allowed to flow in the DC side of the bridge, current also flows through the AC side of the bridge. This means gate current is provided at the AC side of the bridge when the phototransistor conducts current on the DC side.

You should also notice from the diagram that once the triac starts conducting, it passes current from terminal 1 to terminal 3, which is called the *output terminal*. The power that passes through the triac also flows to the neon indicator. This indicator is illuminated every time current passes through the triac and is ready to be used to power an output.

A second neon lamp is in parallel with the fuse for the output. The fuse is provided to protect the electronic circuit against over current. If the fuse "blows," or opens, current is shunted through the neon lamp indicating that a fuse has blown. There is not adequate current available to power an output if the fuse blows.

Figure 10-6 shows a diagram of four output circuits placed together in one module. Notice the internal electronic circuitry for the output is indicated by the rectangular box. When an output signal is received at each output, it enables that circuit's triac, and an output current is supplied at the module's output terminal. This should give you an understanding of how the P.C. accepts an input signal and sends an output signal. Some input modules and output modules operate on voltages other than standard 110 volts or 220 volts. These modules can have different electronic components than the circuits just explained, but the operation is very similar.

The next part of this chapter explains how these input modules and output modules are numbered, or addressed, so the processor can find them. These numbers are also required so the technician will connect the input switches and output loads to the proper module terminals.

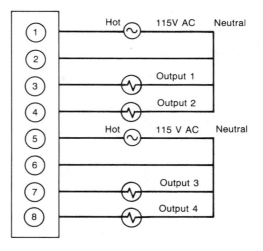

Figure 10-6 Field wiring diagram for 115-volt output module. Diagram courtesy of Gould Modicon Programmable Controller Division.

WHY A NUMBERING SYSTEM IS NEEDED

The input and output circuits just described are grouped together and placed in
modules. Each manufacturer produces a variety of input and output modules. These
modules contain two, four, or eight input or output electronic circuits. Each module
is exactly like the others, having the same voltage rating. Since the programmable
controller may have hundreds of input or output modules, it is important to have
some kind of numbering system to identify each module and all its inputs and
outputs. Since these modules are grouped together and connected in racks, or
housings, an identification system for these is also needed.

Allen-Bradley's smaller systems have modules that each contain four outputs
or inputs. Their larger systems like the PLC/2/15, 2/20, 2/30, and PLC-3 all use
modules that contain eight inputs or outputs. Both type systems group two modules
together for numbering purposes. Figure 10-7 shows this numbering system.

00	10
01	11
02	12
03	13
04	14
05	15
06	16
07	17

Figure 10-7 Terminal numbers for Allen-
Bradley modules. Diagram courtesy of
Allen-Bradley Company.

You should notice that Allen-Bradley uses an octal numbering system to
number its input and output terminals. This means that only digits 0–7 are used.
This is why the first eight terminals are numbered 0–7, and the second eight are
numbered 10–17. These modules are placed together in a rack. The rack provides
the other half of the edge connector for modules to plug into. These racks are
numbered for identification and come in various sizes that hold up to 16 modules.
These racks are pictured in Fig. 10-8.

Figure 10-8 Typical Allen-Bradley I/O racks. Picture courtesy of Allen-Bradley Company.

INPUT AND OUTPUT NUMBERING SYSTEMS

Modicon groups four inputs per module in its 484 system. The Micro 84, the 584, and 884 all group eights inputs per module. Modicon uses a slightly different technique for grouping their modules in housings.

You will remember from chapter 5 that an I/O diagram may be needed to determine exact module and terminal locations (refer to Figure 5-9 in chapter 5). Since Modicon does not use rack or housing numbers in the terminals address, a small dip switch is used to tell the processor in which housing the module is located. This dip switch is located in the top of each housing and is called a *strip select*. The strip select dip switch is set to a 4-bit binary code to identify the housing number.

Modicon has also devised a relatively simple numbering system to identify each input or output terminal on each module and the housing number that these modules are connected to.

It may be easier to understand the importance of the numbers in the address if you realize that the address is used by both the processor and the technician. The address tells the technician where to connect each input switch and output load.

When the modules have problems, the address also directs the technician to the correct part. The processor uses the address to send the electronic signal to turn on or off the proper output and to know which input it has received an input signal from.

Since outputs are grouped with inputs, two columns are addressed together by the processor and are called a *housing*. This means a housing will hold a total of 16 modules. You should also note that four housings are grouped together to make a channel. These four housings will group together a total of 64 modules. Since each module has four inputs or outputs, this allows for 256 inputs and/or outputs to be grouped into one channel. The total P.C. system can then group together up to two channels for each processor.

You should also notice from the diagram in chapter 5 how the system is numbered. Starting in the first column in housing 1, the inputs in the first module are addressed by the processor as 1001 through 1004. The second module is numbered 1005–1008. Remember, each module in the housing has numbered its input circuits as 1 through 4. These numbers are printed on the outside of the module right beside each input neon indicator. Since the processor would get mixed up and confused calling each of its 256 inputs 1 through 4, each input receives a new address or number depending on where it is placed in the housing. You might want to think of this like the addresses on a street in your town or city. All the houses on one side of the street are numbered consecutively. In the case of the first Modicon housing, these numbers range from 1001 through 1032. The modules in housing 2 are numbered 1033 through 1064. This numbering continues until all 125 inputs in channel 1 have been numbered. The numbering continues in channel 2 with 1129 through 1161, each housing uses 32 more numbers or addresses.

Outputs in housing 1 receive similar numbers for addresses except their numbers all start with 0. For instance, the outputs in the first module in housing 1 are numbered 0001 through 0004. These numbers continue through the first channel, just like the numbering for the inputs, through 0128.

You must try to understand that the numbers used in these addresses are very important. They are the only method the programmer and the P.C.'s processor have of keeping track of 128 switches or 128 programmed contacts. This means that if a start button is connected to input 1006, the only way the processor can correctly operate the program is if the programmed contacts have that same number, 1006, as the input switch.

Hopefully you are starting to understand the numbering system that makes the addresses. You may be a bit confused about why each module comes from the factor numbered 1–4 instead of having the full address number such as 1006. Remember, the manufacturer makes thousands of modules and sells them to hundreds of industrial users. Each user only installs the number of modules they will need to operate their system. If each module has an address number 1001 through 1128 marked on it, the user would need to stock 64 different modules. Since all input modules of the same voltage are identical and the address is formulated by where it is placed into the housing, the user only needs to stock one type of input module for each voltage type.

HOW A RACK OF I/Os OPERATE

The Allen-Bradley system controls its rack in a typical manner. The racks are numbered in the octal numbering system. Each rack has an adapter module that is placed in the rack on the far left side (see Fig. 10-9). This adapter helps the processor identify the rack and communicate with each input or output module in the rack. A set of dip switches in the rack is set to provide the rack with its address. Check manufacturers' specifications for setting these addresses.

The processor communicates with the racks through a twisted pair of wires. The twisted pair of wires allows racks to be installed up to several thousand feet from the processor. The twisted pair is used in the serial transmission of data. All signals sent to the rack or received from the rack consist of two parts. The first part is the address. The address simply tells the rack which input or output module

Figure 10-9 I/O rack adapter module. Picture courtesy of Allen-Bradley Company.

the signal is sent to or is coming from. The second part of the signal indicates if the line is high or low. The high signal on an input indicates the input module is energized. A low signal indicates it is not energized. A high signal to the output module energizes its triac to provide an output current. A low signal will turn the input module off.

Each of these signals causes the status indicators to glow and help in troubleshooting. Now that you know how the signals should work for the system to operate correctly, you will be able to look for signals when the system doesn't function correctly.

TYPES OF MODULES

Most P.C. manufacturers produce similar types of input and output modules. The most-used modules are the 110-V. AC input and output modules. Other widely used modules are the DC input and output modules. Some P.C.'s have several DC voltages available in their modules.

There are several other types of modules beside the standard input and output modules. One of these modules is called the *contact output module*. This module, pictured in Fig. 10-10, is made especially for interfaces that need contacts for

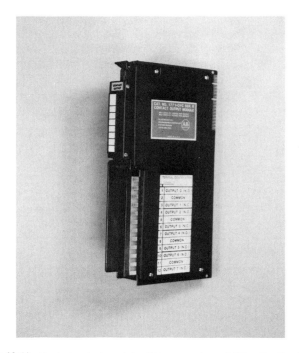

Figure 10-10 Contact output module. Picture courtesy of Allen-Bradley Company.

control, instead of the solid-state devices. This type of output module is needed because contacts can open a circuit and cause current to go to zero, whereas all electronic output modules allow some small leakage current to remain. For some applications, this leak current is unacceptable. Other modules like these include the reed relay modules. All modules provide the P.C. a means of interfacing with the real world, and manufacturers upgrade modules as new applications become known. Use the information in this chapter to answer the questions at the end of the chapter.

QUESTIONS

1. What is an input signal?
2. What is an input module?
3. What is an input device?
4. What is an opto coupler?
5. What is a status indicator?
6. Explain the operation of an input module.
7. Explain the operation of an output module.
8. What does it mean when the input status indicator glows?
9. What does it mean when an output status indicator glows?
10. What does it mean when the blown fuse status indicator glows?
11. Why is a numbering system needed for I/O modules?
12. In which housing and module would you find output 0023? Use Fig. 5-8.
13. In which housing and module would you find input 1083? Use Fig. 5-8.
14. Explain the role of the I/O adapter module in the Allen-Bradley P.C. system.
15. Name several types of input and output modules.
16. Explain the method Allen-Bradley uses to number their I/O racks and modules.

chapter 11
Power Supplies

All programmable controllers need some type of power supply to provide regulated voltage to the processor and other electronic chips in the programmable controller. This regulated power supply is usually low voltage direct current (DC). Since AC (alternating current) is the standard industrial voltage, a power supply's main job is to convert the AC voltage to low voltage DC. Each P.C. manufacturer determines the exact level of DC voltage that its processor and electronic circuits require. Some manufacturers build their power supplies into the cabinet of the processor, while others prefer that you purchase the power supply separately. You will need to understand the operation of the power supply so that you can install it and check it for proper operation.

BASIC POWER SUPPLY OPERATION

A power supply has several jobs to perform. Figure 11-1 shows a diagram of a simple power supply. First, the power supply uses a transformer to lower the 110 or 230 volts AC to approximately 30 volts AC. Next a diode rectifier is used to change the AC to DC. Finally, a filter and regulator are used to provide clean low voltage DC power. These three parts are basic to all power supplies, and their individual operation is easy to understand.

The transformer used to lower the voltage in the power supply is wired as a step-down transformer. This means that the transformer has more turns in its primary winding than its secondary winding (see Fig. 11-2). The transformer's turns ratio depends on what value of secondary voltage is required. Some power supplies need

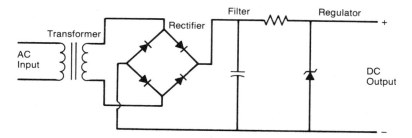

Figure 11-1 Basic power supply.

30 volts, while others need only 12–15 volts on the secondary side. The transformer operates on the principle of induction. This means that there are no moving parts in the transformer, just the two coils. Therefore if voltage is applied to the primary coil, voltage should be present at the secondary coil. If voltage is not present at the secondary side, then the transformer must be bad. When a transformer fails, it usually has an open circuit in one of its coils.

The transformer is usually easy to find in the power supply. It is also easy to test if the power supply is not providing any output voltage. Some P.C. manufacturers allow you to repair your own power supplies, such as transformer replacement, while other manufacturers suggest replacement of the complete power supply.

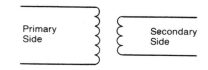

Figure 11-2 Transformer turns ratio.

POWER SUPPLIES RECTIFIER SECTION

The rectifier section of the power supply changes the low voltage AC that comes from the transformer secondary to DC voltage. Most power supplies use diodes to rectify the AC voltage. The power supply will generally use four diodes as a bridge rectifier to provide full-wave rectification. Full-wave rectification allows the power supply to have less ripple and smoother DC power going into the filter section.

The rectifier section of the power supply can also be easily tested. If the power supply is not producing any DC output voltage, you should first test the transformer for correct input and output voltage. If output voltage is present at the secondary side of the transformer but no DC output voltage is present, then the rectifier section should be tested.

An oscilloscope should be used to test the rectifier section of the power supply. The oscilloscope should be connected across the input section of the rectifier where

Figure 11-3 AC wave form as seen on oscilloscope.

an AC wave should be present. Figure 11-3 shows what the AC wave should look like. One note of caution should be observed at this time. Since your power supply is connected to 110v. or 230v. AC, you need to identify all "hot" wires in the power supply. Never connect the oscilloscope ground terminal to any hot wire, because a short circuit will occur in the power supply and cause damage.

The output of the full-wave bridge rectifier should look like the signal in Fig. 11-4. If no voltage is present or if the voltage is too low, one of the diodes in the rectifier is bad. In this case the output DC voltage would be unusable. The power supply rectifier would need to be changed, or the whole power supply should be replaced.

Figure 11-4 DC wave forms as seen on oscilloscope.

FILTER AND REGULATOR

After the signal comes from the full-wave rectifier it is sent to the filter and regulator part of the power supply. The main components of this part of the power supply are the capacitor and voltage regulator.

The capacitor is used to remove ripple from the output signal. Figure 11-5 shows the DC as it appears when it is unfiltered. Figure 11-6 shows the resultant filtering of the full-wave DC signal. The small signal that is the result of filtering is called *ripple*. The ripple for a regulated power supply is usually less than 1 percent, so the output signal is relatively pure DC. It is very important for the DC signal to be as pure as possible because the peak of each ripple may be interpreted as a pulse in a digital circuits.

A voltage regulator circuit is the final part of the system. A voltage regulator circuit is used to keep the output voltage at a constant level. It helps to understand why a voltage regulator is needed if you examine line voltage in any industry. The

Figure 11-5 Unfiltered DC.

Figure 11-6 Filtered DC.

line voltage in most factories is supplied at approximately 208 or 480 three phase voltage. If 110 volts is needed for P.C.'s, it is derived from a step-down transformer. The line voltage usually fluctuates 5 to 20 volts during any hour of the day.

These fluctuations are caused by motors and other large loads being energized and deenergized during normal operation. If you watch the voltage supplied to a motor when it starts, you will notice it drops 5 to 10 percent for just a few seconds. When the motor or large load turns off the voltage may increase 5 to 10 percent for a few seconds. These increases and decreases in voltage cannot be tolerated in the programmable controller processor. Remember the difference between a high and low digital signal may only be three to four volts.

The voltage regulator is usually contained in an electronic chip and keeps the output voltage constant within a small percentage regardless of fluctuations in the input voltage.

If you notice that a processor seems to operate erratically, you should check the output voltage from the power supply with an oscilloscope to see that the output ripple and voltage regulation are within limits. Since the problem may be intermittent, you may need to leave a strip recorder connected to the power supply for several days. Sometimes if the power supply and processor are not grounded correctly, output voltage from the power supply may become erratic. In this case, a proper grounding system, including a grounding rod at the P.C. and an isolation transformer, may be required to eliminate the problem.

TYPES OF POWER SUPPLIES

Programmable controllers use one of several types of power supplies. Some systems, like Modicon, integrate the power supply into the main processor cabinet. This means that the power supply is housed in the same case as the P.C.'s processor (see Fig. 11-7). It is usually mounted on a printed circuit board and is easily removed and replaced if a problem is found. Other systems, such as Allen-Bradley, have power supplies that are not housed in the processor. Instead the power supply is a separate component that is usually mounted near the processor (see Fig. 11-8).

If the Allen-Bradley system has several racks, an additional power supply may be required. In this case the additional power supply may be mounted with the remote racks over 1000 feet away from the main processor. The power supply may also be mounted with the modules in the rack and be capable of being locked out for maintenance and control. See Fig. 11-9.

You should remember the power supply is a small but very important part of the P.C. system. It generally operates correctly and quietly during the life of the system. In fact the power supply usually is the most reliable part of the system. If you do have problems with the P.C. system that point to the power supply, you now have enough understanding of its operation to use the manufacturer's instructions to test it. Several questions regarding the power supplies operation are presented at the end of this chapter. Use these questions to test your knowledge.

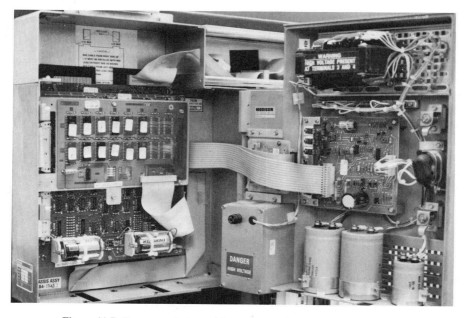

Figure 11-7 Power supply housed in processor cabinet. Picture courtesy of Gould Modicon Programmable Controller Division.

Figure 11-8 Stand alone power supply. Picture courtesy of Allen-Bradley Company.

Figure 11-9 Power supply mounted in rack with keylock. Picture courtesy Gould Modicon Programmable Controller Division.

QUESTIONS

1. What purpose does the power supply serve in a P.C. system?
2. Explain the operation of the power supply. Be sure to include all four sections.
3. Explain how a transformer operates.
4. How would you test a transformer?
5. Explain the operation of the diode rectifier section.
6. Why is a rectifier section required?
7. Explain how you would test the diode section using an oscilloscope.
8. Show the input and output signal that you see when you test the rectifier section.
9. Explain the operation of the filter section and why it is needed.
10. Explain why a problem could result in shorting out the power supply if you are not careful when using an oscilloscope to test the transformer or rectifier section.
11. Explain why a voltage regulator is needed as part of the power supply.
12. Name two places you would look to find the power supply in a P.C. system.
13. When should you suspect the P.C. power supply of having a problem or fault?

chapter 12
Processor
Operation

THE PROCESSOR

The heart and brains of a programmable controller is its processor. The processor, as you know, is a very powerful computer, housed in a chip. The chip is mounted on the main printed circuit board and may have several auxillary chips connected to it. These chips aid in the storing of memory and control input/output operations.

Figure 12-1 shows a typical processor and memory mounted on a printed circuit board. Notice that the printed circuit board allows all the chips to be connected to each other without wires. Instead, the electronic signal moves along the lines of foil on the printed circuit board. If you must touch a printed circuit board for any reason, do so only at the edges because static from your hands can get into the circuit and possibly cause damage if you touch any conducting foil. You must also be sure to protect the printed circuit board from cracking while you remove it, because a crack in the conducting foil acts as an opening in the circuit.

Some processors have extra empty chip sockets provided in case you want to increase the memory size. Other manufacturers, like Allen-Bradley, connect their memory with printed circuit board edge connectors. If you must add extra memory, be sure to work carefully around the main printed circuit board by first removing all electrical power. The processor is typically housed inside an enclosure. Figures 12-2, 12-3, and 12-4 show typical Allen-Bradley, Texas Instruments, and Modicon processors.

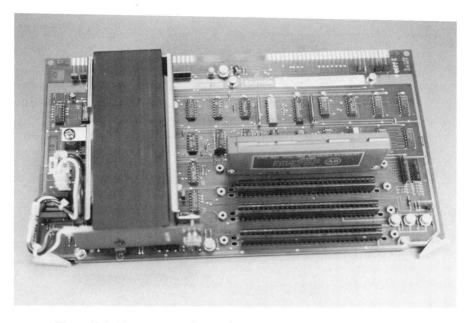

Figure 12-1 Memory mounted on a printed circuit board. Picture courtesy of Allen-Bradley Company.

Figure 12-2 Typical Allen-Bradley processor. Picture courtesy of Allen-Bradley Company.

Figure 12-3 Typical Texas Instruments processor. Picture courtesy of Texas Instruments.

Figure 12-4 Typical Modicon processor. Picture courtesy of Gould Modicon Programmable Controller Division.

HOW THE PROCESSOR OPERATES

The operation of the processor is very complex. You could spend several years in electronics classes just to begin understanding its operation. What is important about programmable controllers is that the user and programmer only need to understand a few facts about its operation to have a successful P.C. system.

The processor has a basic system operation, or master program, permanently stored in its memory. This program is needed to get the P.C. system to respond when you turn it on. In other words, if there were absolutely no program in the processor when you connected the industrial terminal with the CRT and energized the system, nothing would happen. Therefore the master program in the processor does several things. First, it energizes the system, tells the processor to send a start-up menu to the CRT, and opens a channel to receive signals from the keyboard. This enables the operator to see the menu on the CRT and begin making selections to start programming on the keyboard. Other operations from the processor's basic internal program complete the routing and control of all incoming signals from the keyboard. This master program is usually called the *executive*, or *ROM program*. The executive program also monitors a master switch on the P.C. panel. This switch is changed by the programmer or technician to indicate whether the processor uses a program keyed into memory or operates a program stored in memory. Some P.C.'s also have a third selection that allows the processor to test a program in memory without energizing outputs.

Allen-Bradley calls the positions on their master switch *Run Program*, *Run*, *Test*, and *Program*. Modicon uses menu selections instead of a switch to set up their system. Once the selection is made, the processor begins running that portion of its executive program.

If the technician selects the program mode, the processor recognizes signals from the keyboard and begins building the program. If the technician selects the run or test mode, the processor begins to execute the user's program and looks for input signals and sends output signals. If the technician wishes to stop the program but leave the P.C. system powered up, the *stop processor* selection can be made.

The same executive program in the processor also provides the user a means to change the way memory is configured.

MEMORY LAYOUT

The memory operation in all processors is slightly different. Since each manufacturer uses the processor chip of their choice, each system is slightly different. These differences are usually in the size of storage space for programs, counters, timers, messages, and other registers. Figure 12-5 shows a typical memory layout.

You can see that the memory is divided into parts. Each part can contain a different amount of data. Since data is made up of bits that are on or off, the memory is mapped out according to how many bits it can handle. Since 16 bits

Total Words

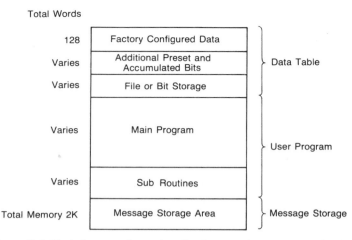

128	Factory Configured Data	
Varies	Additional Preset and Accumulated Bits	Data Table
Varies	File or Bit Storage	
Varies	Main Program	User Program
Varies	Sub Routines	
Total Memory 2K	Message Storage Area	Message Storage

Figure 12-5 Typical memory layout for Allen-Bradley. Picture courtesy of Allen-Bradley Company.

are grouped together to form *words*, you can see from this figure that Allen-Bradley memory is described in terms of the number of data words it can handle.

Starting at the top of the memory map you can see that factory-configured data uses 128 words of memory. Notice that counters, timers, and sequencers store preset and accumulative values along with other file functions in the same area. The inputs and outputs used in the program are stored as bits in the bit/word storage area. Each input or output uses one bit. This area is called the *Data Table*. Since the total number of counters and timers can be adjusted slightly, the total size of this Data Table area also varies.

The next section of memory is reserved to store the main program. The user determines the size of main programs. Any unused memory space can be then designated for use by other functions, like timers, and counter or file functions. Likewise if only a few timers are needed, the user can designate the unused timer, counter area for other uses, such as an enlarged program or messages. This enables the user to have flexibility in using the P.C. memory. The user can also designate part of the user program for a subroutine. The exact amount of subroutine space again can be selected and adjusted, depending on the size of the main program. The main program and subroutine area is designated as a user program area.

The last area of memory is called *message store area*. This part of memory can also be shortened or enlarged to meet the needs of the system. If the P.C. system is used for complex process control, the message area can be used to store various messages pertaining to operating conditions. If more or less message area is needed, the memory can be configured to accommodate this.

The ability to change memory configuration is not available on all processors.

Some systems provide a set number of counters and timers. If you do not use them, the registers area becomes wasted space in memory. If more counters or timers are needed, there is no way to get them other than buying an extra processor. If the processor does allow memory to be configured, the changes can be made through a few menu selections.

CONFIGURING P.C. MEMORY

Once you have an idea of how many counters, timers, and files you will need and how large the main program will be, you can configure the processor's memory. Some processors allow the user to complete the configuration with the industrial terminal and keyboard.

Other P.C. systems do not have on-board ability to configure the system. Instead you actually determine the P.C. configuration when you purchase the system. Since you know approximately how much memory and how many counters, timers, inputs and outputs you need, you can buy the size that will handle this. Since the majority of P.C. systems are mainly input and output controllers, memory configuration has not been a real problem. Now that P.C.'s are being used for advanced automated control and process control with messages, the ability to configure memory is nearly a necessity.

MEMORY SIZE

Some other P.C. systems get past the problem of configuration by allowing memory to be expanded as the process grows. This means that a system can be originally purchased with a small memory to control one machine with a few inputs, outputs, and timers. Later the system can be enlarged by adding several memory segments to control hundreds of inputs, outputs, and timers. Some P.C. manufacturers feel this is a better way to approach system expansion, because the user does not need to understand the memory table configuration. If you need more timers or program space, just buy more memory. It seems that as competition increases among P.C. manufacturers a variety of configuration methods will be provided.

Presently, you can specify the size of memory in thousands of bytes. Each one thousand bytes is called *1K of memory*. The K is the abbreviation for the word *kilo*, which means one thousand. Typical memory sizes are 1K, 2K, 4K, 8K, 16K, 32K, 64K, and 128K. It is likely that within a few years megabyte memory will be fast, reliable, inexpensive, and small enough to be used routinely in programmable controllers. The word *mega* denotes that memory contains millions of bytes. Once the program is stored in memory, the processor must execute it.

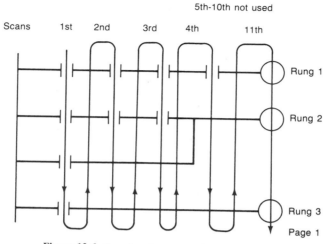

Figure 12-6 Scanning the program by columns.

PROGRAM SCAN

When the processor checks memory and starts to execute the program, it is called *scanning the program*. All processors scan the program in one fashion or another. Some begin at the top left part of the program and work through the program by columns. Figure 12-6 shows this operation. Other programs are scanned rung by rung. Most P.C.s show a format for the number of columns or elements that can be in each rung. For instance, Modicon 484 can support ten elements plus one output in each rung or program line. This restriction is determined by the processor. Since the processor is going to scan eleven columns, the number of elements in each line cannot make the program wider than eleven columns. Some processors can scan an unlimited number of columns, but they only display a specific number of columns at one time on the CRT.

    The processor views the program as one long page, even though it only displays a limited number of rungs on the CRT. The program is mainly broken into pages or sections for the user.

    Even the printer tends to print the program by rungs, sections, or pages. The processor, though, scans the whole program, column by column or rung by rung. The speed of the scan is generally very fast compared to human standards. It varies from as fast as four milliseconds to several hundred milliseconds, depending on memory size and other variables. Remember, a millisecond is one thousandth of one second.

    This means that each rung of the program is scanned many times in one second, and inputs, outputs, timers, counters and math registers are updated quite often. In simple P.C. systems, scan time is rarely considered. But in larger P.C.

systems used for process control or complex automation, the scan time becomes an important issue. In some large programs, the logic is so complex the scan time actually varies depending on what's happening to certain inputs, outputs, and comparison functions. When this happens, clocks and math functions sometimes are not updated often enough, and they become slightly inaccurate. If this occurs, the programmers and technicians must check the consequences to the total system. If the inaccuracies affect the machine operation, parts of the program can be skipped from time to time or subroutines can be made to make the critical functions more accurate. Faster processors and better understanding of critical and noncritical functions will keep this type of problem to a minimum in the future. Some systems are also offering a real-time clock in the module that operates with a high degree of accuracy independently of the processor. These clock modules fit into the racks and housings just like other modules.

Once you understand the operation and configuration of memory, you can use this knowledge to set up your programmable controller for your specific application much easier. When the system is programmed and configured for optimum operation, major problems with memory size and scan time are eliminated. You need to be patient and try several solutions when problems with scanning arises. Remember the computer or processor is still relatively "dumb." It takes a skilled technician to make the P.C. system function correctly.

Several questions are provided at the end of this chapter to increase your knowledge of this area.

QUESTIONS

1. Explain what the processor does in the P.C. system.
2. Explain what the executive, or ROM, program does for the P.C. system.
3. Show a brief diagram of a typical memory layout.
4. Explain how memory may be divided into parts for a data table, user program, and message storage area, and why the size of these sections is variable.
5. What is meant by the term *1K of memory*?
6. Explain two ways memory size may be varied.
7. Explain two methods the processor could use to scan a program it is executing.
8. Identify some problems you might encounter if the program scan becomes too slow.

chapter 13
Alternative Programs

Some models of programmable controllers, like the Texas Instruments 5 TI and GE Series One and Series Three P.C.'s use a type of programming called *gate logic*. Logic gates are the basic building blocks of these processors and their programs. These gates include AND gates, OR gates, and NOT gates. The NOT gate is also combined with AND gates and OR gates to form NAND gates and NOR gates. Programmable controllers that use this type of program are less expensive to manufacture and can be designed to control up to 20 inputs and outputs.

AND GATES

The AND logic gate is one of the simplest logic gates to understand. The most basic AND gate logic circuit has two inputs and one output (see Fig. 13-1). The AND gate circuit provides an output only when both inputs are present. If either input signal is missing or "low," then the output will not be present. When a signal is present, it is said to be "high" and is represented by a 1. If the signal is not present, it is said to be "low" and is represented by a 0. You may remember that 1 and 0 are the two values of the binary numbering system.

Another way to explain the operation of logic gates is with a truth table. A truth table simply lists all the possible input signal conditions and the resulting

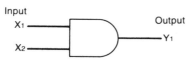

Figure 13-1 AND gate logic symbol.

In X_1	In X_2	Output Y_1
0	0	0
1	0	0
0	1	0
1	1	1

Figure 13-2 Truth table for two-input AND gate.

outputs. The truth table uses 1s and 0s to show if a signal is high or low. The truth table for the AND gate is shown in Fig. 13-2. From the truth table, you can see that the AND gate's terminals are listed as input X_1, input X_2, and output Y_1. The truth table is always explained in terms of its inputs and the resulting effect the inputs have on the output signal.

In the case of the AND gate, there are four possible conditions for the two inputs. These conditions are: both X_1 and X_2 off; X_1 on and X_2 off; X_1 off and X_2 on; both X_1 and X_2 on. Remember the terms *high*, *on*, and 1 all mean the same, and *low*, *off*, and 0 mean the same thing.

From the AND gate truth table you can see the only condition that provides an output for the AND gate is when input X_1 and input X_2 are both high.

Another way to explain the AND gate concept is with switches or contacts. From Fig. 13-3, you can see that the contacts in an AND gate circuit are actually in series with each other. You now can see why both input X_1 and input X_2 must be on or closed to get the output signal. The AND gate logic program in the P.C. looks very similar to the diagram in Fig. 13-3. To put this program into this type of programmable controller's memory, three program statements would be required.

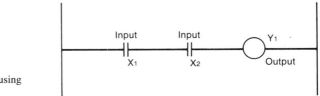

Figure 13-3 AND logic circuit using contacts.

From Fig. 13-4 you can see that the first statement in each line is a store statement. This means that you want to put a set of open contacts into the rung against the power rail and start a new rung. The AND X_2 means you want X_2 in series with X_1. OUT Y_1 means you want output Y_1 to come on if the logic is true. Other logic gates will be explained in this chapter, and sample programs will be presented for you to try at the end of this chapter. Some systems simply use only numbers such as 1, 2, or 3 for their inputs and outputs rather than use X and Y.

Store X_1	or	Store 1
And X_2		And 2
Out Y_1		Out 1

Figure 13-4 Program for AND gate circuit.

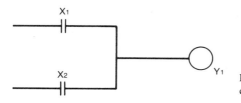

Figure 13-5 OR logic circuit using contacts.

OR GATE

The OR gate is another basic logic gate used in computers and programmable controllers. Figure 13-5 shows two contacts in an OR gate circuit with an output. The operation of this circuit shows that if either input X_1 or input X_2 is closed, a signal is provided to the output. You can see the OR circuit is a simple set of parallel contacts.

The logic symbol and truth table for the two input OR gates is shown in Fig. 13-6. By looking at the parallel contact circuit representing the OR gate in Fig. 13-5 you can interpret the truth table very easily. The OR gate provides an output when either input 1 or input 2 is high, or when both 1 and 2 are high. Notice the only time the output is low is when both 1 and 2 are low.

You might have figured out that the names of the logic gates describe their operation. For instance the AND gate requires input 1 (and) input 2 to be high for the output to be high. For the OR gate to have an output you would need a high signal at input 1 (or) input 2.

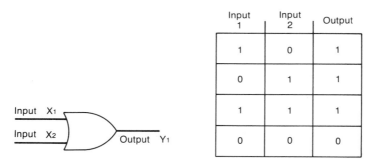

Input 1	Input 2	Output
1	0	1
0	1	1
1	1	1
0	0	0

Figure 13-6 OR gate symbol and truth table.

MULTIPLE INPUT AND OR GATES

You may also come across AND gates and OR gates that have more than two inputs. For instance, the circuit in Fig. 13-7 shows four sets of contacts connected in series with one output. Since series contacts represent an AND gate, this circuit

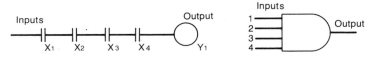

Figure 13-7 Four-input AND gate circuit and symbol.

represents a four-input AND gate. The diagram for the logic symbol is also shown in Fig. 13-7. The logic is still the same. Input 1, input 2, input 3, and input 4 must all be high for the output to become energized.

An OR gate could also have more than two inputs. From Fig. 13-8 you can see that the four-input OR gate has four sets of parallel contacts. The OR logic produces an output if either input 1, input 2, input 3, or input 4 is high.

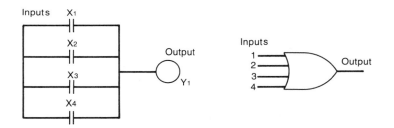

Figure 13-8 Four-input OR gate circuit and symbol.

You may also find any number of AND gates and OR gates in the same circuit. Figure 13-9 shows a sample circuit that has both AND gate and OR gate inputs. The best way to understand this circuit is to treat every contact that is in parallel as OR gate logic and every contact that is in series as AND gate logic. Therefore, you must use OR gate logic rules for parallel contacts and use AND gate logic rules for series contacts. In Fig. 13-9 you can see that both contacts X_3 and X_4 must be closed at the same time as contact X_1 or X_2 to get the output energized. This approach also makes programming this circuit rather simple. Figure 13-10 shows the program statements for the circuit in Fig. 13-9.

As you progress to more complex circuits, you can understand them more easily if you break them down to their AND gate and OR gate logic level.

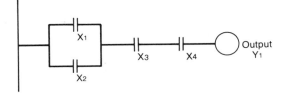

Figure 13-9 AND gate and OR gate circuit.

Store X1
OR X2
AND X3
AND X4   **Figure 13-10** Program for circuit in
Out Y1   Figure 13-9.

NOT GATE

The last major building block in the logic family is the NOT gate. The NOT gate is sometimes called an *inverter*. The easiest way to understand a NOT gate is to think of a set of normally closed contacts operated by a coil. If a signal is (not) present at the coil, the contacts pass power. If a signal is present at the coil, then the contacts would open and (not) pass power.

From Fig. 13-11 you can see that the NOT gate, or inverter, has only two terminals, input and output. If power is present at the NOT gate input, then no power is available at the output.

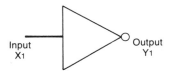

Input
X1

Output
Y1

Figure 13-11 NOT gate symbol.

This means that a high, or 1, at the input gives a low, or 0, at the output. If a low, or 0, is present at the input, then a high, or 1, is present at the output. From the truth table in Fig. 13-12 you can see that whatever signal is present at the input, the opposite signal is present at the output. This means that the input signal is always inverted to get the output signal. Hence the name, inverter.

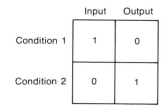

	Input	Output
Condition 1	1	0
Condition 2	0	1

Figure 13-12 NOT gate truth table.

Figure 13-13 shows the NOT gate as a circuit diagram and in program form. The program circuit symbol for the NOT gate is very similar to the logic symbol in that the normally closed contacts have one input on the left side and one output on the right side. You can assume the NOT function is being used any time you see the normally closed contact symbol in the program circuit.

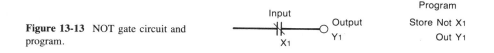

Figure 13-13 NOT gate circuit and program.

NAND GATES

The NOT gate can also be combined directly with the AND gate and the OR gate. Figure 13-14 shows the logic symbol circuit and truth table for the AND gate with the NOT function connected directly.

Input 1	Input 2	Output
1	0	1
0	1	1
1	1	0
0	0	1

Figure 13-14 NAND gate symbol and truth table.

The small circle on the output terminal of the AND gate logic symbol shows that the output should be inverted. This new logic gate is called *NAND* gate for the NOT AND logic function. From the circuit diagram and truth table in Fig. 13-14 you can see that the only time the output would receive a signal is if contact X_1 and contact X_2 are both low, or closed. A program for the circuit in Fig. 13-14 is presented in Fig. 13-15.

You will notice the first contacts always go into the program with a store instruction. Since the first component is normally closed contacts, the NOT function is used. The NAND function is the next instruction. It is used to show that the NOT X_1 contacts and the NOT X_2 contacts are in series. The Out Y_1 instruction sends the output signal to output 1 when the NAND logic is true. Practice programs are provided at the end of this chapter. You may need some practice before this programming becomes easy.

Figure 13-15 NAND gate program and circuit.

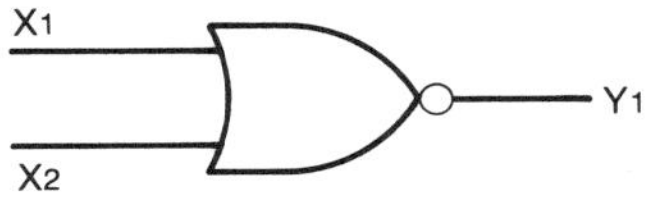

X₁	X₂	Y₁
1	1	0
0	1	0
1	0	0
0	0	1

Figure 13-16 NOR gate symbol and truth table.

NOR GATES

The inverter, or NOT function, can also be connected directly to the output of the OR gate. Figure 13-16 shows the NOT symbol, a little circle, at the output end of the OR gate. This new logic gate is called a *NOR gate*. From the truth table and circuit for the NOR gate shown in Fig. 13-16, you can see that the output is present when either or both input signals are low (0). The only time the output signal is not present is when both input X_1 and input X_2 are high (1).

The program instruction for the circuit in Fig. 13-16 is listed in Fig. 13-17. This program is similar to the NAND gate program. The first set of normally closed contacts is entered into the program with a STORE NOT instruction. The NOR X_2 function shows the processor that you want the two sets of normally closed contacts in parallel. Other programs using NOR instructions are provided at the end of this chapter.

Store NOT X₁
NOR X₂
OUT Y₁

Figure 13-17 Program for NOR gate circuit.

TIMERS AND COUNTERS

The smaller programmable controllers also provide timer and counter functions. These smaller P.C.'s are widely used in the plastics industry. The timers and counters allow plastic presses to have variable timers and cycles. These timers and counters are very similar to the timers and counters used in larger P.C.'s.

The timers and counters are programmed into a set of 10 or 20 registers that are dedicated, or set aside, for this purpose. This means that if your programmable controller has 20 counter/timer registers, you can assign the register any way you like. For instance, you may need 12 timers and 8 counters, or you may need 16 timers and 4 counters. You have twenty registers to use as either counters or timers as you desire. Some machines may need only timers. In this case you could assign all twenty registers as timers and have no counters.

These twenty registers are numbered sequentially from 1 to 20. If you have used register 1 as a timer, T-1, you cannot use register 1 as a counter. Therefore, if you have timer T-1, you cannot have a counter C-1. The timer could be T-1, and the counter C-2. In other words you cannot have a timer and counter with the same number.

The actual timer operation and programming format is very similar to timers in larger P.C.'s. These timers provide time delay on and time delay off.

TIMER FORMAT

The timer is identified on the top line of the timer function block. From Fig. 13-18 you can see this is TMR1, or timer 1. The next line indicates timer preset value. The preset time is set in $\frac{1}{10}$-second increments. The bottom line in the timer block sets the accumulative, or current timer, value.

The timer function block has two terminals on the left side and one output terminal on the right side. The top left terminal is the control, or enable, terminal. The bottom left terminal is the reset terminal. The timer begins timing only when the reset and control lines are energized. Any time the control line is opened, the timer stops accumulating time. When it is closed again, the accumulative time begins again. Anytime the reset line is opened, the accumulative, or current, value returns to 0. Anytime the accumulative, or current, value equals the preset value, the output is energized. The output coil can control a number of sets of normally open and normally closed contacts to provide a time delay on and time delay off function.

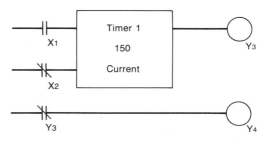

Figure 13-18 Typical timer.

TIMER OPERATION

Figure 13-18 shows a program using timer 1. The preset value is set for 15 seconds (remember, preset time is in $\frac{1}{10}$ seconds). The timer begins timing when the X_1 enable contacts close. The reset contacts X_2 must also be closed. Output Y_4 is energized as soon as the X_1 enable contacts are closed and will delay off after 15 seconds. You may need a set of X_1 contacts in series with the normally closed Y_3 contacts to keep Y_4 off until X_1 enables the timer. The Y_3 output is energized delay on after 15 seconds. This timer is constructed or programmed as a retentive timer. If you needed a nonretentive timer, the X_1 enable contact would also be used for the reset contacts. In this way any time contacts X_1 opened, the timer would be reset.

If you need the timer to automatically reset, the reset is controlled by output contacts Y_2. Basically this timer provides the same operating functions as the timers found on larger P.C.'s. These timers can also be cascaded with other timers and counters to provide long time delays.

COUNTERS

The counters provided on these smaller P.C.'s operate similar to the counters on larger P.C.'s. Figure 13-19 shows the format of a typical counter. The function block shows the CTR number on the top line. The middle line of the function block shows the preset value or current count value. The current value constantly updates the total counts in the counter. You can examine the total count anytime by viewing the current value. The counter function block has two input terminals on the left side and one output on the right side. The top terminal is the enable, or count terminal. Every time this line is energized the accumulative, or current, count value is increased by 1 until the preset value equals the accumulative value. When these two values are equal, the counter is full, and the output is energized. The reset terminal must be energized for the counter to add values to the accumulative current value. When the reset line is deenergized, the accumulative value returns to 0. As you can see, this operation is very similar to counters on other P.C.'s.

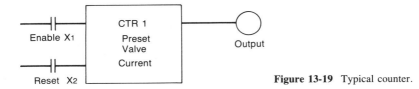

Figure 13-19 Typical counter.

COUNTER OPERATION

Figure 13-20 shows counter 4. This counter has a preset value of 8 and a current value of 0. The counter adds 1 to the current value each time X_1 contacts close. The reset contacts X_2 must stay closed, or energized, during this time. When the current value is equal to 8, the output Y_1 is energized. If larger counts are required, the counters can be cascaded.

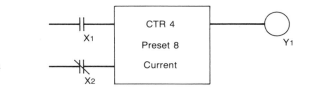

Figure 13-20 Counter as seen in a
program.

If you need the counter to automatically reset, the Y_1 output contacts would be used for the reset operation. You can see these operations are similar to the counters found in larger P.C.'s.

The smaller P.C.'s provide a very useful function. Machines that only need a few inputs, output timers, and counters are suited for these small P.C.'s. They also tend to be very economical, yet they provide diagnostic capabilities just like the larger machines. Many technicians consider these systems as relay replacers. They can do this and much more.

QUESTIONS

1. Draw the symbol for a two-input AND gate.
2. Draw a circuit that uses three contacts and one output for an AND circuit.
3. Draw the symbol for a two-input OR gate.
4. Draw a circuit that uses three contacts and one output for an OR circuit.
5. Draw the symbol for a NOT gate.
6. Draw a circuit using a NOT function.
7. Draw a symbol for a NAND gate.
8. Draw a symbol for a NOR gate.
9. Draw a circuit that includes two NAND gates, two NOR gates and one output.
10. Draw a diagram of a timer, and explain the control, reset terminal, and the output.
11. Explain how you would diagram a retentive timer.
12. Explain how you would diagram a nonretentive timer.
13. If the largest preset value of your timer is 999, draw a diagram using a set of timers that provide a 200-second time delay to output Y_3 (remember preset time is in $\frac{1}{10}$ seconds).
14. Explain how the counter operates. Include the preset value, current value, enable input, reset input, and output.
15. If the largest preset value is 999, show the diagram of a set of counters that could count values up to 25000.

chapter 14
Installation

At some time you may become involved with the installation of a P.C. system. The installation can be easy if you understand several critical points. First, the inputs and outputs should be grouped according to actual system location. This means all I/O modules for a specific machine should be installed in the same rack or housing or cabinet. Other groupings should be made to make sure that similar type modules are kept together. Lastly, all modules that use the same voltage should be grouped together when possible.

DETERMINING THE NUMBER OF I/O MODULES

First, you should lay out a diagram of the system that you intend to control. All inputs and outputs should be identified. These identifications should include the type of I/O, such as regular input, output, TTL, or Analog. The voltage requirement of each I/O should then be determined. Then location of each machine input switch and output should also be determined and noted on the diagram. When all of these things have been identified, a parts list can be made and an advanced diagram can be developed to show actual rack location in each electrical cabinet. After the racks are identified, the modules can be laid out, and actual wiring diagrams can be developed.

The wires that connect switches and coils to the modules should be identified to show the address of the I/O. A master list of inputs and outputs should be made, and address numbers should be applied. Figure 14-1 shows a typical list of racks

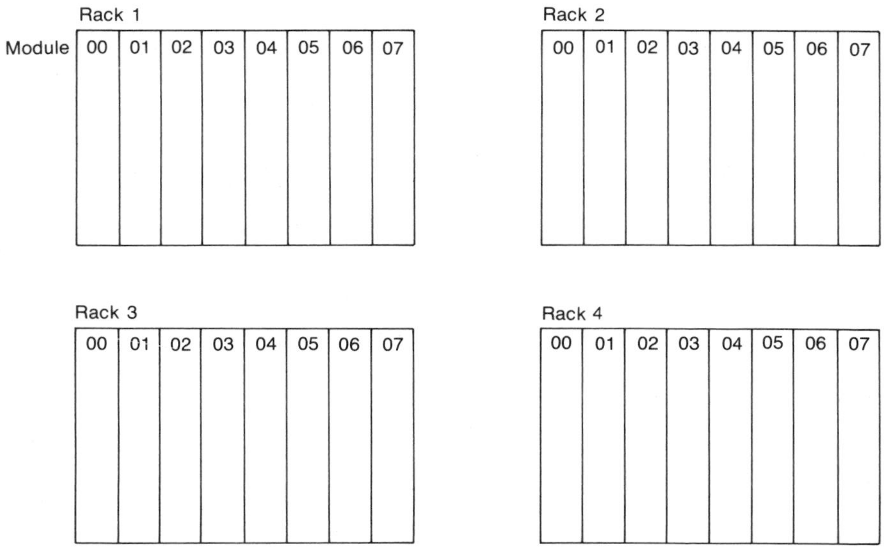

Rack 1

Module	00	01	02	03	04	05	06	07

Rack 2

00	01	02	03	04	05	06	07

Rack 3

00	01	02	03	04	05	06	07

Rack 4

00	01	02	03	04	05	06	07

Figure 14-1 Typical rack identification. Diagram courtesy of Allen-Bradley Company.

with addresses ready to be filled in. This list should be kept as a permanent record that is called *I/O documentation* and used for troubleshooting.

FIELD WIRING

Once you have completed the diagrams, you can begin to mount the components in the electrical cabinet. Be sure to follow the manufacturer's specifications on air space around each component, such as the processor, to ensure proper ventilation. Also be sure to measure the ambient air around the installation. Extra provisions for cooling cabinets may be needed at some locations, like foundries.

A word of caution when you are installing P.C. components in the cabinet. You must remember to mount components, like the processor, power supply, and modules, where you can see their fault indicators. You must also remember that these components must be mounted in such a way that they can be easily removed and replaced.

Once you are ready to wire the modules, you should take time to label every wire that goes to a module terminal with the terminal's address. This way, when you are troubleshooting or if modifications are made, you can easily trace out the wires.

The actual field wiring should be placed in some type of cable tray inside the cabinet so you can get to all wires. The actual connections at the module end should be made with field wiring arms, or terminals. Try to avoid wiring directly to modules, where possible. If you are installing a small P.C. you will probably end up wiring directly to the module's terminals. The reason field wiring arms are used is to make module replacement as simple as possible. If field wiring arms are used, you will be able to exchange modules without any changes in wiring. Some P.C. manufacturers do not give you a choice anymore. The only way to wire to their modules is through field wiring arms.

At the machine end of the system, all wiring must be installed strictly by the National Electric Code. Your P.C. manufacturer may provide you with some additional considerations, such as not putting wires with different voltage signals, like TTL and 220 AC, in the same conduit because this may cause interference to the TTL signal.

During the wiring procedure, be sure to identify all machine inputs and outputs two ways. First, each component should be identified with its machine name, such as LS1 for limit switch number one. Next you should be sure the switch is also identified with its P.C. address. This address should also match the number on the wire that runs to the device, as well as match the number of the terminal and module. You will also recall this same number is used to identify all contacts in the program that this device controls.

This may sound like a lot of extra work, but you must remember that someday this P.C.-controlled machine is going to malfunction. If every component is clearly identified, troubleshooting and repairs should take only a short time. If components are not properly identified, you may search for hours for a simple broken wire or bad limit switch. Remember, the total system may have thousands of these switches. Figure 14-2 shows a typical cross-reference sheet for a system. This sheet should be developed as the installation takes place. When the machine is in operation this sheet should be kept in the cabinet for referral during troubleshooting.

When the field wiring is installed, all P.C. cables should also be installed. Most of these cables are already made by the manufacturer. All you need to do is snap them into the proper terminals. Also be sure that all racks and processor dip switches are set for proper identification. Remember, since there may be several

Switch/Coil	P.C. Address	Machine Location	Program Rungs
LS1	111/02	Ram Up	10, 16, 22
LS2	111/05	Ram Down	12, 14, 18
Start	111/01	Start/Stop Station	1, 6,
MS1	010/02	Motor Starter	3

Figure 14-2 Typical cross reference sheet.

Dip
Switches

Figure 14-3 DIP switches used for strip select in Modicon 484 housings. Picture
courtesy of Gould Modicon Programmable Controller Division.

hundred modules, each rack or housing must be identified. Figure 14-3 shows a
picture of setting the housing identification for a Modicon 484. Be sure to check
the manufacturer's specifications and procedures for setting these switches. If more
than one processor is used in a master-slave set up, each processor will also need
to be identified with its internal dip switches.

Once all wiring has been completed, be sure to take time to retrace all routings
and connections. One mistake here can be very costly. Remember, you are still
working with a computer.

POWER SUPPLIES

During the planning process, you indicated specific voltage and current requirements
for all parts of the system. You should be aware that not only does the processor
need a power supply, but so will the modules in the racks. In fact, if you have
TTL modules, analog modules, and other advanced modules, you may need four
or five separate power supplies. Each one will have a specific voltage range, and
their size must match the amount of current needed. Don't be afraid to double check
the number of supplies needed. Sometimes these items are overlooked during the
planning stages and cause project cost overruns when they must be purchased later.

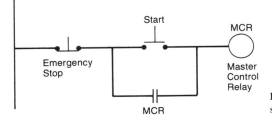

Figure 14-4 Example of a hardwired safety circuit.

SAFETY CIRCUITS

You should be aware that a hardwired safety circuit should be built into the main power circuit for the I/O modules. Figure 14-4 shows an example of this type of circuit. From this example, you can see the safety circuit is a simple stop-start switch connected to the master control relay. The contacts of this relay control the power to the main transformers and power supplies for the I/O modules.

This hardwired safety circuit is designed to provide emergency stop capabilities if something should go wrong in the P.C. Keep in mind, the only difference between an output being turned on or off in the P.C. is the difference between a 1 and a 0 in the data word. It is possible for the computer, or processor, in the P.C. to have a failure and inadvertently turn on some outputs. In this case, the emergency stop button could be used to remove all power to the output module at that time. Additional stop buttons may be added to this circuit if needed. They would be placed in series with other stop buttons just like other electromechanical stop switches.

You must also check your local code and shop safety procedures to make sure your system is safely installed.

GROUNDING

Some P.C. systems tend to have strange problems. The actual problem can not really be pinpointed. In fact, the system may seem to have a little gremlin inside it. In these problems, the system can malfunction erratically and lose partial memory. Generally these problems can be traced to poor or faulty system ground. To correct this problem, extra grounding rods should be placed in the ground right at the P.C. cabinet. Be sure you use the grounding procedures your P.C. manufacturer provides. If you have strange processor problems, don't accuse the system of being haunted. Rather, make a complete check of the system's ground.

Complete the questions at the end of this chapter using the new knowledge you have gained.

QUESTIONS

1. Name the steps that should be completed during the installation procedure.
2. Explain why a field wiring diagram must be prepared prior to installation.
3. Explain why the components, such as limit switch, should have both a machine identification and a P.C. address.
4. Explain the function of field wiring arms.
5. Explain several considerations that should be made in determining the placement of P.C. components in the electrical cabinet.
6. Explain why you should go to all the trouble of making a cross-reference sheet for your installation.
7. Explain how it is possible to have three or four different power supplies in the P.C. system.
8. Explain the purpose of the MCR safety circuit. How does it differ from the MCR instruction found in some P.C.'s?

chapter 15
Troubleshooting

One of the main advantages of using a programmable controller is having its troubleshooting capabilities. Since all switches are connected to an individual input and all output loads are connected to individual outputs, their status can be continually monitored. The only problem in using the P.C. for troubleshooting is that most technicians do not realize how to use the system to troubleshoot itself. This chapter explains some of the troubleshooting aids available on P.C. systems.

STATUS INDICATORS

The most obvious troubleshooting aid available on P.C.'s is the status indicator. As the name implies, the status indicator lamps show the status of the components in the system. Each critical part of the system usually has some type of status indicator. By checking these indicators a technician can usually determine the location of the problem.

When the system has a problem, the technician must make an initial test to determine how much of the system is disabled, in some manner. If the processor or racks are involved, the fault is usually deemed a major fault. If the fault disables a single input or output, it is generally considered a minor fault.

The major fault indicators on most systems show when a major portion of the P.C. system or components have failed. These fault indicators monitor the input voltage, power supply voltage, processor run or stop mode, and communications between racks and processors. Figure 15-1 shows the typical location of these status

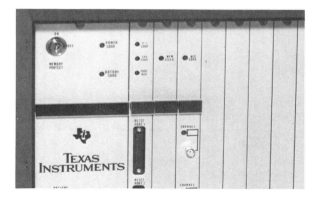

Figure 15-1 Location of status indicators on the Texas Instruments Model 560/565 processor. Picture courtesy of Texas Instruments.

indicators on the processor. The power supplies usually have red status indicators that glow when the input voltage or regulated output voltage has a fault. Some processors use a green indicator to show when the system is okay, while others use the absence of the red indicator as the signal that the system is operating correctly. The processor usually has one or more lamps to indicate the status of the processor. Most of these indicators are LEDs (light-emitting diodes) and have a long operational life. One of these LEDs shows the operational mode of the processor. These modes include run, stop, and program. Other LEDs indicate processor fault, memory faults, or other processor problems. The third major part of the system that has status indicators is the adapter module. The adapter module is located in the I/O rack, and its status indicator glows when communications between the rack and processor are functioning correctly. When the indicator is not illuminated, it indicates that a communications fault has occurred.

The technician will have some idea of the magnitude of the problem when a fault occurs. By checking the status indicators on the power supply, processor, and adapter modules most major faults can be located. Usually if one of these indicators shows a fault has occurred, the troubleshooting process can then be concentrated in that area. If no major indicators are showing, then the troubleshooting process will be more extensive.

One last point about processor status indicators. The LED that indicates the processor's mode should be checked any time the system has a problem. This indicator is usually overlooked for a much larger problem. Many times the only problem in the system is that the processor has stopped for some reason. Some processors automatically return to the stop condition any time a disturbance or fault has occurred. Many hours have been spent by excellent troubleshooting technicians in search of complex electronic problems when all that has happened is that the processor has stopped.

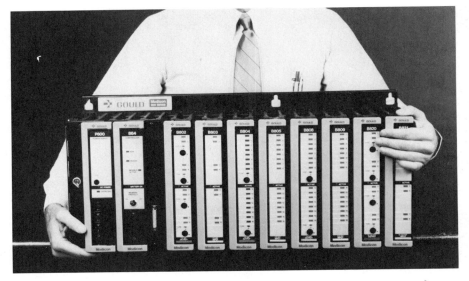

Figure 15-2 Location of status indicators on Gould modules. Picture courtesy of Gould Modicon Programmable Controller Division.

I/O STATUS INDICATORS

Each programmable controller manufacturer uses some type of indicator to show the status of input and output modules. Figure 15-2 shows the typical placement of input and output module status indicators.

The typical status indicator on the input module is energized any time an electrical impulse is received at the input terminal. This occurs when the input switch, or contact, has closed. Each individual input terminal has its own status indicator.

If a fault is isolated to a point where the technician suspects the input to the P.C., this side of the circuit can easily be checked. Figure 15-3 shows the typical components connected to the input side of the system.

From the diagram, you can see that all a technician would need to check for is the presence of the input status indicator. If the indicator lamp was illuminated you would know that the input switch and connecting wire were operating correctly. As you open and close the input switch, the status lamp should go off and on. If the lamp fails to turn on, the switch and wire should be checked for loss of power or an open circuit. If the indicator lamp comes on but does not go off, the switch and wire should be checked for a short circuit or welded contacts. This technique is probably the most important point in the troubleshooting method. Since only one switch and one set of wires are connected to each input terminal, the problem should be easy to trace when the status indicator shows a fault.

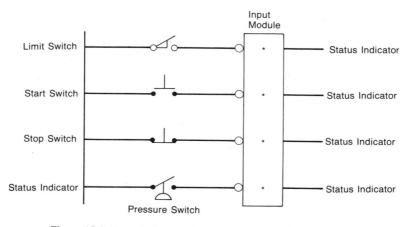

Figure 15-3 Status indicators show whether a switch is open or closed.

In some older electromechanical systems, several switches are connected in series to control an output. When the output shows a circuit failure the technician must test or inspect each switch and interconnecting wires in that circuit to find the problem. In comparison the P.C. system has only one switch per input with one wire. If a fault is detected, the technician concentrates the troubleshooting effort to that switch and wire. Even if the technician is inexperienced, the system can be fixed by replacing each of the three components until the fault disappears. You should also remember that Allen-Bradley uses these same indicators for its examine on and examine off instructions. If you are using Allen-Bradley P.C. literature, you will notice that the normally open contact is called *Examine On*, and the normally closed contact is called *Examine Off*. This Examine On instruction means that if the status indicator is on, the logic in the program is true and the normally open contacts pass power. The Examine Off instruction means that the contacts pass power only when the status indicator is in the off condition.

OUTPUT MODULE STATUS INDICATORS

The output module has two types of status indicator. One indicator shows when an output is present at the module, and the other indicator shows if an output fuse has blown. The main output indicator shows when the processor has sent a signal to the output terminal. Each terminal on the output module has its own indicator. If the module indicator is illuminated, the module should have voltage present at the output terminal. If the indicator is not illuminated, then no voltage is present at the terminal.

The blown fuse indicator in the output module is a little different for each manufacturer. Even some models made by the same manufacturer are different. Usually, there is one blown fuse indicator for each terminal in the module or one

Figure 15-4 Replacing a blown fuse. Picture courtesy of Gould Modicon
Programmable Controller Division.

indicator for the whole module. If there is only one indicator for the whole module
you generally know which fuse is bad by checking for the output load that does
not operate. Figure 15-4 shows a fuse being replaced in a module. In fact most
troubleshooting is started when an output load, such as a motor, fails to run.

Using the status indicator on modules and other parts of the system, faults
can be isolated rather rapidly.

USING THE CRT AND PROGRAMMING PANEL
TO TROUBLESHOOT

Another part of the system that is designed to aid in troubleshooting the P.C. system
is the CRT and programming panel. The keyboard provides several special function
keys that make troubleshooting rather simple. The CRT combines with these functions to highlight devices that are energized and show if devices are on or off.

The easiest of these functions to use is the CRT's ability to highlight rungs
or devices that are energized. Basically if you check the suspected rung of programming or the input and output devices, the CRT highlights those that are energized.

Another similar function, the GET function in the Modicon P.C., allows you to call up a specific input, output, or register in the lower portion of the CRT and display its current condition. If it is energized, the word *ON* appears. If it is not energized, the word *OFF* appears. If it is a counter, timer, or math function the current accumulative value for the register is displayed. If you have a Modicon P190 programming panel, it can be used to display an additional page showing I/O and register status.

If you are looking for a problem with I/O or a specific value in a counter, timer, or register, this function enables you to closely inspect the condition of these devices.

BIT EXAMINATION

The Allen-Bradley systems have a similar feature. All timers and counters show the accumulated time right below the device in the rung where it was programmed. If you need to inspect the condition of an input or output device or the contents of a register, the Allen-Bradley system provides functions called *data display*. By displaying the word address of a specific module, each output or input in the module word address is displayed, and each individual bit of the word can be examined. Figure 15-5 shows an example of the word address displayed. Any 0 displayed in a bit location indicates the input or output is deenergized, and a 1 indicates the device is energized. The hardest part for someone using the bit examination, or data display, is determining which bit is used for each device. This can become easier if you remember that the memory address and module address are identical. It will also help to remember that bit 0 is at the far right of the address and the numbers increase to the left to bit 7 in the first eight bits. The second eight bits start at bit 10 and increase to the left to bit 17. These bit numbers correspond to the terminal numbers on the input and output modules. The same examination technique can also be used to check the contents of specific math functions and other data in memory. Each function and piece of data is given a memory address where the processor stores the data. By using the industrial terminal and a CRT to show the data display of the word address, you can display the contents, or data, stored there. You should remember that this data may be displayed as a binary number or as a BCD, a binary-coded decimal value. Be sure to check the instruction provided in the manufacturer's documentation for their format.

Word III	17	16	15	14	13	12	11	10	7	6	5	4	3	2	1	0
Data	0	1	1	0	1	0	0	1	1	0	0	1	0	1	1	1

Figure 15-5 Data display of an input word (16 bits) indicating which inputs are on or off. Diagram courtesy of Allen-Bradley Company.

Once you have become proficient in checking the condition of a bit in an I/O address or the values in a data address, notice how much easier troubleshooting can become. You will also be amazed at just how much data the processor manages.

SEARCH FUNCTION

Since you know how to use the highlight function on the CRT, you will find it difficult to look through hundreds of rungs for the specific input or output that is not functioning properly. Most programming panels have a function key called *SEARCH*. The SEARCH function essentially looks through the program to find each place that a specific input, output, or register has been programmed. For instance, if the motor starter that controls a conveyor will not come on, you should look up the address of the output module, and use the SEARCH function. If you are using a Modicon system and the output address is 0025, you key in 0025, and use the SEARCH function. The CRT then displays the whole line of program connected to output 0025 (see Fig. 15-6). The CRT's cursor would be over top of output 0025. You can then use the GET function, highlight, or bit examination to see if 0025 is energized.

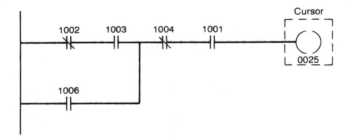

Figure 15-6 Typical program line displayed by the search function on the CRT.

If the output is not energized you then use the GET function or bit examination to see if these inputs are energized. Since the output is not energized, one or both of the inputs must be deenergized. By using combinations of these functions, the troubleshooting actually becomes entertaining as you use the CRT, program, and status indicators to look for clues. The actual operation of the SEARCH function is slightly different for each P.C. system. Allen-Bradley systems format requires that you use the exact program symbol, then the address, then the SEARCH function key. Figure 15-7 shows the instructions that are entered from the keyboard to search the program for input 111/02. If input 111/02 is used in more than one rung in the

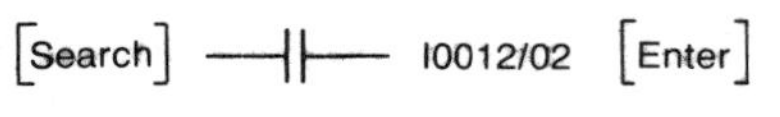

[Search] ——| |—— I0012/02 [Enter]

Figure 15-7 Typical Allen-Bradley PLC-3 search function. Diagram courtesy of Allen-Bradley Company.

program, the continue key is used until the CRT indicates the last occurrence. If the input 111/02 was never programmed, then the CRT indicates the search has failed.

On Modicon Micro 84 a numbered code has been devised to indicate which type of device you intend to search. From Fig. 15-8 you can see that if you key in 0 search, 0026 search, the processor looks for output 0026. If you key in 1 search, 1025 search, the processor scans for input 1025. The larger Modicon 484 and 584 do not use this number scheme. Their search function is similar to Allen-Bradley where only the address is needed. The Modicon 484 and 584 can also search for one input used on several different program lines by using the CONTINUE SEARCH key on the keyboard. When the last occurrence of the device has been displayed the CRT indicates the search is complete.

You can see that the SEARCH function is probably the most powerful and useful function for troubleshooting large programs. In fact, when you discuss programmable controllers with technicians who work on large automated systems they will tell you the system would be impossible to troubleshoot and manage without a P.C. and the SEARCH function.

Figure 15-8 Modicon Micro 84 search functions. Diagram courtesy of Gould Modicon Programmable Controller Division.

0 means search output coil or contacts

1 means search input contacts

3 means search any register

4 means search any device

FORCE AND DISABLE

The last major troubleshooting function that is available on most P.C. systems is called a *FORCE function*. The DISABLE function is the companion function of force, but its operation is the opposite of the FORCE function. The FORCE function allows the technician to select an input or output and FORCE its condition to on or off. Some P.C.'s allow only the outputs to be forced, while others allow both inputs and outputs to be forced.

It must be emphasized at this time that only the most skilled and experienced technician should use the FORCE function on an automated system because the FORCE function can be used to override or bypass all safety features built into the program. For instance, if your system uses a heating loop or furnace, the heating element can be forced on for testing. But this FORCE function passes all temperature safety devices. If the force is left in place, damage would surely result because the safety devices have been bypassed. Another problem can arise when the operation of a machine is sequenced. By forcing a specific input or output, you may cause major machine motion out of sequence and damage the machine or injure someone working on the machine.

The FORCE function is provided to energize or deenergize a component in the system so it can be more easily tested during troubleshooting. A good example of using the FORCE function is to troubleshoot an output that only stays energized for 1–2 seconds in the normal program. Since this does not allow enough time to check for the problem, the FORCE function can be used to keep the output energized until the technician can get to the device and test it. But the technician at the keyboard must analyze the total machine operation to be sure that energizing this output for an extended period will not injure anyone or cause damage to the machine.

It must be pointed out that the program must be carefully inspected to make sure that all FORCE and DISABLE functions have been removed before the P.C. is returned to normal automatic operation. This is especially true if safety devices were disabled to all outputs that will be energized during troubleshooting. If a low-pressure safety was disabled for troubleshooting, it must be returned to normal condition for automatic operation, or the machine will operate without a low-pressure safety circuit and damage could occur.

Each P.C. manufacturer uses a slightly different format for forcing and disabling a device in the program. You should check with your system instructions and allow someone to help you try these instructions out. Remember, *never randomly force or disable inputs and outputs in an active program, because severe damage could occur to the machine or cause personal injury!*

DEVELOPING A TROUBLESHOOTING TECHNIQUE

Now that you understand the ways the programmable controller is able to troubleshoot itself, you need to develop a technique or system that isolates the problem as quickly as possible.

The most efficient way to troubleshoot a P.C. system is to start at the output side. Usually an operator or supervisor notices the system is not operating correctly and asks for a technician to check out the system. Since the outputs cause the actual system action, it is usually the lack of action that is noticed.

Figure 15-9 shows the diagram of a system with a typical malfunction. It is a motor that drives a conveyor. The motor is started with a motor starter, and the motor starter is controlled by three switches. The operator notices that the conveyor has stopped and requests a technician to check out the system.

The technician starts checking the system at the output side. The easiest test to make is inspection of the status indicator. It may be necessary to look at the cross-reference sheet to see the actual program and module address of the motor starter for the conveyor. In this example, the output's address is 010/03. The technician looks to rack 1, module group 0, and checks if the status indicator for terminal 03 is illuminated.

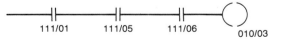

Figure 15-9 Typical system program with a malfunction. Diagram courtesy of Allen-Bradley Company.

If the status indicator for output 03 is illuminated, the technician checks for voltage at the module terminal and motor starter. If voltage is present at the module terminal 03 but not at the motor starter, the interconnecting wire is suspect. If voltage is present at the motor starter, then the problem is in the motor starter's coil, contacts, or with the motor itself. It may seem overly simplified, but if the module status indicator is illuminated, the problem must be between the output module and the motor. This makes the troubleshooting task much easier.

If the status indicator for the output module is not illuminated the technician revises the troubleshooting plan to intensify the search in the program between the processor and output modules. Since the status indicator for output 010/03 is not on, the next check uses the CRT to see if output 010/0/3 is highlighted in the program.

If the output is highlighted but the modules status indicator is not illuminated, the problem is located between the processor and the rack. Typical problems here would be an inoperative output module, a bad terminal on the output module, an inoperative adapter module, a bad rack or wires between rack and processor. Each of these areas could be checked easily.

If output 010/03 is not highlighted in the program the emphasis of the trouble-shooting shifts to examining the contacts that enable output 010/03. Since contacts 111/01, 111/05, and 111/06 must all be energized for the rung logic be true and the output energized, each of these contacts should be examined. Either the highlight function or bit examination function is used to check these inputs. If the bit examination function is used the technician uses the data display to display data word 111. This data word shows bits 00 through 07. Bits 01, 05, and 06 are examined for 1 to see if they are energized. If the CRT highlight function is used, the inputs are checked to see if each contact is highlighted. If input 111/06 is not highlighted but the other two are, the test shifts to the input section. The module status indicator on input terminal 111/06 is checked to see if it is illuminated. If it is illuminated but the contacts are not highlighted in the program, then the problem is between the input module and the processor. These problems would typically be the inoperative input rack, bad adapter module, bad interconnecting wires or the processor chip itself. These components are easily checked to find the problem.

If the status indicator is not illuminated, then the problem is narrowed down to the input switch, the input modules' power supply, or the input module itself. The testing is now moved to the input area of the P.C. system. Voltage at the input module is the easiest to check. If voltage is not present, the power supply, or power source, is tested next. If power is present, the problem has to be in either a broken wire between the switch and module, or bad contacts in the switch. The contacts are usually suspected and tested first.

You have just seen an example of a troubleshooting technique that leads you to the problem every time. This method is shown in diagram form in Fig. 15-10 and quickly leads you to any problem in the system. It is assumed that the technician uses a voltmeter and ohmmeter to test for the presence of voltage and for open circuit in wires and switches.

Troubleshooting Chart

1. Is output load operational? If yes, go to step 18. If no, go to step 2.

2. Check Output Status Indicator. If off, go to step 7. If on, go to step 3.

3. Check for proper voltage at output module. If no voltage, go to step 4. If volts OK, go to step 5.

4. Check power supply to module. If no voltage, replace power supply and go to step 18.

5. Check voltage at output load. If no voltage, check and repair bad wire. If voltage is present, go to step 6.

6. Check output load for continuity and proper operation. Replace load, and go to step 18.

7. Check program for highlighted output. If highlighted, go to step 8. If not, go to step 9.

8. If problem is between processor and output module, Check out module, I/O adapter, connecting wire, and processor. Change as necessary. Go to step 18.

9. Check all contacts which enable output in step 8. If all are highlighted, go to step 10. If all are not highlighted, go to step 11.

10. Check to see that the processor is in RUN mode. Check or replace processor if output is still not highlighted.

11. Check unhighlighted contacts. If contact number indicates it is controlled by input switch, go to step 13. If contact number indicates it's controlled by memory coil, go to step 12.

12. Check the program line where the memory coil is listed (use search function). Is memory coil energized for correct operation? Remember, about normally closed contacts. If problem is found, go to step 18. You may find a "chain event" where some contact is keeping the memory coil deenergized. Check the contacts, and follow thorugh using the instruction in step 11.

13. Check Input Status Indicator. If on, go to step 14. If off, go to step 15.

14. Check module connection to rack, check rack, check I/O adapter, and check interconnecting wire or processor. Replace as needed. Go to step 18.

15. Check voltage to module. If correct voltage is present, go to step 16. If correct voltage is not present, go to step 17.

16. Check for loss of voltage to module. Replace transformer, power supply, or fuses as needed, then go to step 18.

17. Check input contact and wiring repair, or replace as needed, and go to step 18.

18. When problem is corrected, test system for proper output operation. If output is still inoperative, return to step 1 for additional problems. If output functions correctly, you have successfully troubleshot and repaired the system. Return system to normal operation.

Figure 15-10 Troubleshooting chart.

One of the main reasons for using a programmable controller to control automated systems or process control is its ability to test each part of the system and show its current status. If the troubleshooting technician understands all the troubleshooting functions on the P.C. and uses a systematic approach to problem solving, downtime is kept to a minimum, and most faults are found quickly.

QUESTIONS

1. What is a status indicator?
2. Explain why the processor, power supply, and modules may all have status indicators.
3. What is an LED?
4. Name two status indicators that can be located on the processor.
5. Name two status indicators that can be located on an I/O module.
6. Explain the term *examine on* as used to indicate a normally open contact on an Allen-Bradley P.C. system.
7. Explain the term *examine off* as used to indicate a normally closed contact on an Allen-Bradley P.C. system.
8. Explain why blown fuse indicators are used on output modules.
9. Explain how the HIGHLIGHT function on the CRT can be used to help you troubleshoot.
10. Explain how the BIT EXAMINATION function is called up on the CRT and how it can be used to troubleshoot.
11. Explain why the Allen-Bradley data display does not include outputs 8 and 9 (refer to Fig. 15-5).
12. Explain how the SEARCH function operates.
13. Explain why you may have to search for more than one instance of a contact in a program.
14. Explain how the FORCE and DISABLE functions can be used.
15. Explain why it is very dangerous to use the FORCE and DISABLE functions on an operating system.
16. Explain why the troubleshooting technique should always check the output as a first test.
17. Use the troubleshooting guide listed in Fig. 15-10 to locate problems in a typical P.C. system. Be sure to follow all safety procedures.

chapter 16
Maintenance

Most industrial users of programmable controllers are beginning to see the advantages of using the P.C. system to monitor periodic maintenance of motors and other machines. It is very important to ensure that motors and other machines are periodically lubricated and adjusted at proper intervals. It is a proven fact that a proper periodic maintenance schedule helps prevent costly downtime repairs and helps extend the working life of the machine. The only problem has been that it is very difficult to monitor the exact running time of each motor and machine so that time between lubrication and adjustments is not too long or too short. If the periodic maintenance is completed too often, it becomes too expensive.

If the time interval is too long, machines tend to break down more often and use extra energy. This means that if a machine's exact running time is recorded, a maintenance schedule can be planned to provide the lubrication and adjustments at the optimum time. This maintenance must also be scheduled around machine production and machine failure that causes downtime. The P.C. can be utilized to keep track of running time and provide printed documentation for periodic maintenance schedules.

USING TIMERS AND COUNTERS TO RECORD OPERATION TIME

Most programs in programmable controllers do not utilize all the timers and counters that are provided in memory. In most programs, the scan time of the processor is not increased too much by adding several counters and timers for maintenance.

156

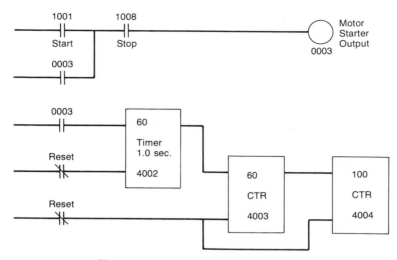

Figure 16-1 Sample timer for maintenance.

These maintenance timers and counters can be programmed during the initial machine installation or added after the machine has been operating for several years.

The simplest way to record machine or motor running time is to use the programmed output for the motor starter to also control the enable contacts for a timer. Figure 16-1 shows this simple program. The top rung of the program has the output 0003 that turns the motor starter on and off. Contacts 0003 are also used to enable the timer and are closed every time the motor starter coil is energized. They remain closed until the motor starter coil is deenergized. This causes the timer to keep track of the actual time the motor is running. The timer is programmed as a retentive timer so it accumulates 100 hours of the motor's running time.

You should notice that the timer preset value is not large enough to hold the value 360,000, which is the number of seconds in 100 hours. The timer is cascaded to two counters to get the full value of 360,000. The first timer times 60 seconds and automatically resets. Every 60 seconds, the timer pulses the output, which enables counter 4003. Counter 4003 has a preset value of 60 for the number of minutes in one hour. Each time counter 4003 counts to sixty, it pulses its output. The output of counter 4003 energizes the enable terminal for the second counter 4004. This counter is actually counting hours because it receives its enable signal once each hour of run time. This requires that counter 4004 has a preset of 100 to count up to 100 hours. At the end of 100 hours total running time, the second counter 4004 sends a signal to latch output 0006. This latched output energizes a warning lamp indicating 100 hours has elapsed and periodic maintenance is required. Since a latch coil is used, manual reset is required to unlatch the coil and turn off the lamp. The manual reset causes a maintenance person to come to the equipment to push the reset. The programmer has several options at this point besides the warning lamp. Some larger P.C.'s, such as the Allen-Bradley PLC-2 family and

PLC-3, allows the output to activate a message that is stored in memory. This message is entered into the program by the programmer and is sent to a printer any time the timer indicates, so a hard copy or printed copy is available. In this case, the message would be stated something like this: "THE MOTOR ON CONVEYOR NUMBER 6 HAS ACCUMULATED 100 HOURS OF RUNNING TIME. PERIODIC LUBRICATION, BELT INSPECTION, AND TENSIONING IS DUE." When this message appears on the printer, the maintenance supervisor knows it is time to schedule periodic maintenance. Some companies even have the P.C. message include a work order and work order number so that the maintenance technician can start the maintenance procedure as soon as it is convenient.

Another type of timer is provided on some programmable controllers. It is called a *clock/calendar*. This clock and calendar is adjusted to real time once the P.C. is operational. This means that at any given time the clock/calendar shows the exact day, month, year, and correct time. This function can be used to record downtime by printing the time and date when the machine fails and when it becomes operational again. This helps determine the amount of downtime the machine has, and it gives some type of time profile when the downtime occurs. This can be used in trying to find the causes of machine failure or mean time between failure. In each of these examples, the P.C. can be used to record and store this vital information.

IDENTIFYING SYSTEM PROBLEMS WITH THE P.C.

Some industrial processes have become very complex. In some factories the programmable controller may be controlling a complete production line that has dozens of machines operating together to make a product. In this situation the finished product of one machine may be the raw material to the next machine. This presents a chain reaction type problem when any one of the machines stops. To prevent unwanted outages or down time, the P.C. can be used to monitor all the vital information pertaining to each machine. Included in this information can be motor voltages, currents, and temperature, along with critical machine motions.

The program shown in Fig. 16-2 is taking an analog signal for the voltage, current, and temperature from one motor. It is also monitoring for machine motion. Each parameter is compared to a set point. For instance the over current set point may be set at 5 amps, and the under current set point may be set at 1 amp for this motor. Any time the operating motor current gets larger than 5 amps or smaller than 1 amp, an unsafe condition is indicated, and a signal can trigger a hard-copy message telling the maintenance technician which motor is having a problem prior to the machine failing. This type of program enables the P.C. to warn the maintenance personnel of conditions that could cause a machine failure if it is not corrected immediately.

By combining several of the conditions, a good technician can troubleshoot the system from the warning messages. For instance, if the voltage signal is normal, current is too low, and no machine motion is detected, the technician may suspect

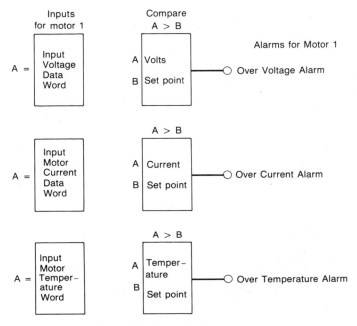

Figure 16-2 Typical maintenance program.

a loose or broken drive belt. Another example could be a loss of voltage, current, and machine motion, which could be a sign of a blown fuse or power loss. You can see that with a little imagination you can write programs so that the P.C. can monitor and alert maintenance personnel of problems even before they cause major outages and long periods of downtime.

TESTING INPUT AND OUTPUT MODULES

In most industrial P.C. systems, the input and output modules are the part of the system that is most often suspected of being at fault when the machine stops. As electricians phased in their first programmable controllers, they were used to swapping parts to see if they were bad. This meant they would swap the suspect part, like a limit switch, with a new one. If the machine started to operate correctly, they solved the problem. On most systems, this was a fast way to get the machine back into commission.

When programmable controllers were first being used in industry, the input and output modules were the part of the system that failed most often. This meant that the input and output modules had to be capable of being changed with a minimum of work. Most systems now have a field wiring arm or wiring terminals that attach to the module with an edge card connector. This means modules can be

Figure 16-3 Easy module removal. Picture courtesy of Gould Modicon
Programmable Controller Division.

removed without disconnecting any wires. This saves time and reduces the possibility of miswiring the new part. These provisions enable the electrician or technician
working on the P.C. to easily swap the module without major problems. Figure
16-3 shows this ease of module replacement.

When you are working on a P.C. and feel that one or more inputs or outputs
are bad on a module, the fastest way to find out is to swap that module with one
that is identical. *A note of caution here!* Be sure you are using an exact replacement
for the module you are swapping, because major damage to the module and P.C.
system results if you try installing a different module than you removed. Manufacturers have used several methods to prevent this from happening. One way to
identify the module is by color codes and numbers. Be sure the modules that you
are swapping are the same color and have the same identification numbers. Another
way to keep from installing a wrong module is to put "keys" on the edge connector
board in the rack.

A key is a piece of plastic that blocks off one of the slots in the edge connector
board so that only modules that are the same can fit into that rack. Keys for each
rack must be put in place by you, the user, once the system is installed. Once keys
are installed, it is important to remember not to try to force modules that are different
into the edge connector because the key or edge connectors on the module will
break.

You must keep in mind that module swapping is a very useful minor maintenance practice, but remember that modules contain sophisticated electronics, and they must be handled with care.

USING THE P.C. SYSTEM AS A RECORDER

Another minor maintenance technique can be utilized on your P.C. system. This technique involves using the counters, timers, and data registers as a recorder. Many industrial applications experience electrical problems called *faults* that occur rather randomly. Usually this fault does not last very long and is very difficult to isolate. An example of this kind of fault might be a set of dirty electrical contacts in a limit switch. These contacts are just dirty enough to operate correctly most of the time, but once in a while the dirt causes the switch to be inoperative. This fault causes the machine to shut down periodically. When a technician comes to check the contacts, they seem to operate correctly again. The machine operates correctly for several more days and then has the fault again. Since there are several limit switches on this machine that could shut it off if they had a fault, it becomes very difficult for the technician to know where to start. Usually when this type of fault occurs, the maintenance technician begins changing switches at random in hopes of getting the right one.

If the machine is controlled by a programmable controller a short program can be written to monitor all the switches. This program would not disturb the program that operates the machine. Instead, it would be written at the bottom of the original program and take advantage of the P.C.'s ability to program a set of input contacts as many times as needed. Since each limit switch is connected to an input, it has its own number. All you need to do for the new monitor program is use these same numbers when you program the test program contacts.

The monitor program would program each limit switch contact to a counter, as in Fig. 16-4. The counters' preset value would be higher than the largest number of times the limit switch would normally be expected to open and close in a week. You may need to cascade several counters for very large numbers. In this example, limit switch 1 is connected to input 1003, and limit switch 2 is connected to input 1007. If there were twenty limit switches on the machine you would program each one to its own counter. Next you would need to analyze the machine's operation to see the sequencer these switches open and close. For instance, in this example, let's say switch 1003 opens and closes two times for every time 1007 opens and closes once. This means that if you checked the counters, 1003's count should always be exactly twice as large as the counter controlled by 1007. The next time the machine shuts down because of a limit switch fault, the technician could check the accumulative count in each counter. If the counter controlled by 1003 had 200 counts and the counter controlled by 1007 had 98 counts, you could assume that switch 1007 is causing the fault. Be sure to account for count differences due to when the cycle was interrupted.

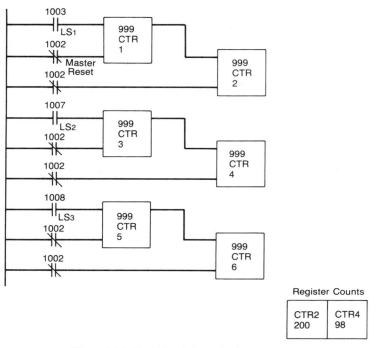

Figure 16-4 Typical switch monitoring program.

You could use the P.C. to record the switches operation even in machines that have hundreds of limit switches. In fact, as the number of switches get larger, this technique would save more time.

Some P.C.'s, like Allen-Bradley systems, provide a contact histogram. This histogram can be programmed to provide a detailed history of the on/off transition of any contacts. This history is maintained and updated in the P.C. memory and allows the troubleshooter to call up the histogram on the CRT or print a hard-copy printout to use when a fault is being checked.

The important point to remember at this time is that once a switch is connected to the P.C., it can be used in as many rungs and subprograms as your processor's memory will allow. This should allow you to program as many maintenance programs as you need to maintain proper machine operation. The only condition to watch for is programs where the scan time becomes critical. If scan time becomes a problem, you could still use the maintenance programs only when faults are suspected, then have the processor skip that portion of the program, or simply remove those rungs that are part of the maintenance program after the fault is cleared.

Once you begin using the programmable controller to help monitor maintenance, you will realize its time-saving and record-keeping abilities are as useful as its ability to be reprogrammed. In fact, once most technicians begin relying on the

P.C. for minor maintenance, it becomes difficult to work on machines with electro-mechanical controls. This should be your goal, too. You should gain the confidence needed to program your system for operation and maintenance.

QUESTIONS

1. Explain how the processor may be used to keep track of a motor's running time.
2. Show a sample program of a retentive timer used to keep track of total motor run time.
3. Explain why counters are used with the timers in the program in Fig. 16-1.
4. Explain why a latch coil is used as the output in the program in Fig. 16-1.
5. Explain why a printed message may be included in the program in Fig. 16-1.
6. Explain how motor voltage, current, and temperature might be used in a program to monitor the motor's safe operation.
7. Explain how you could use the indications from the alarms in the program in Fig. 16-2 to tell what is wrong with the motor.
8. What problems should you be aware of if you try to change out or swap modules?
9. Explain what purpose a "key" serves when installed on the edge connector of a module.
10. Explain how a P.C. can be used as a recorder to tell if one limit switch in a set of limit switches on a machine was having an intermittent fault.
11. Explain what a contact histogram is and how it may be used for finding faults in switches on the machine.
12. Write a maintenance program to keep track of 10,000 hours of motor running time.
13. Write a maintenance program to keep track of the number of times a motor has started on a machine.
14. Write a maintenance program that compares the number of times three limit switches will open and close during normal operation. Explain how the program could be used to determine which switch was having an intermittent fault.

chapter 17

Program Documentation

All programmable controllers have the ability of being reprogrammed. If your industrial process changes, all you need to do is update the program with several changes, and the system is operational again. Being reprogrammable is an outstanding feature for P.C.'s, but it also brings a new set of problems to go with it.

The biggest problem is the memory that can be easily changed for reprogramming can also easily be lost or scrambled. The main user's program of a P.C. is stored in the RAM memory of the processor. RAM stands for *random access memory*. The processor can store and retrieve the program in RAM and execute it.

The processor can also easily update counters, timers, and math functions stored in RAM memory as it scans the program. The actual program and memory data are nothing more than logic gates in the RAM that are high or low. This means each gate in memory is energized or deenergized. These patterns make up the code that we call the *program*. If a major power failure occurs and the back up power (battery) is low, all the logic gates go to the deenergized state and the pattern known as the program is lost. Also, if an erroneous spike of current found its way onto the same electrical circuit the processor was using, parts of memory may be altered. In some early P.C.'s, when memory was lost or altered, the program had to be reentered by hand, causing long downtime for the industrial process.

Since the first time P.C.'s were brought into industries, manufacturers knew a reliable method was needed to store a copy of the program. In those earlier days the best and most reliable way to do this was to transfer a copy of the logic pattern (program) onto another chip for storage. This chip was called a *PROM*. PROM stands for *programmable read only memory*.

The PROM chip copies the program logic by setting its own logic gates to highs and lows by electrical charges. This process is called *burning a PROM*. Once charges set the logic gate high or low, the state remains unchanged when power is removed. This means that PROM chips can store a program and not be bothered by some of the problems facing RAM memory. Also since the PROM chips do not require electricity to keep the program stored, they can be kept in an office file drawer or on the main P.C. board for use if the main processor memory is lost.

When the processor memory is lost, the PROM chips can be installed and the program reloaded from the PROM chip to the RAM memory in the processor. Some manufacturers of P.C.'s left room on a circuit board near the processor to keep the PROM chip right in the system. This way if RAM memory is lost the program can be downloaded from PROM to RAM with a minimum of downtime. Industries took advantage of keeping several different machine operations on different PROMs. This enabled machines such as presses to be reprogrammed to run different parts rather easily.

When the program needs to be changed because of process or design changes that occur from year to year, a new program is entered into the processor and then backup copies can be stored again on PROM chips. Since the old program can not be used after the changes are made the PROM chip containing the old program can be erased or stored for future use. Erasing the PROM chips is a very simple process. The PROM chips logic circuit is sensitive to ultraviolet light. A window is provided on top of each PROM chip to allow ultraviolet light in to erase the chip. When the PROM is programmed, a piece of tape or covering is placed over the window to keep out unwanted ultraviolet light. When the chip needs to be erased, the cover is removed from the window, and the chip is placed under a source of ultraviolet light for several minutes. P.C. manufacturers usually supply the PROM burners and erasers.

PROMs are still being used by some processors as backup memory. But as programs become larger and altered more often, burning and erasing PROMs takes too much time. Newer ways to store programs have been developed. These ways are faster and easier to use than PROMs.

EAROM AND EPROM

The biggest drawback in using a PROM chip is its sensitivity to ultraviolet light. If unwanted ultraviolet light reaches the PROM it alters its memory and makes it useless. If you want to change its program you must first erase the PROM under ultraviolet light for several minutes. Sometimes the chip is not erased cleanly and other problems occur.

Newer types of PROMs were developed in the middle 1970s. These PROMs were erased by an electric pulse rather than by ultraviolet light. These chips were called *EAROM* or *EPROM*, meaning electrically alter read only memory. This meant that the PROM could be erased in several seconds, just prior to reprogramming. Usually the EPROM or EAROM was inserted into an edge connector on the processor circuit board and one button provided a signal to erase the old program and another button started the storage of the new program. Some manufacturers used one button on the EAROM and a signal from the processor to erase or store a signal. The biggest advantages of the EAROM chips is that the storage and erasure process is fast, and the chip is safe from stray ultraviolet light. The memory size is still fairly limited in the EAROM, so several other methods of program storage were further developed.

TAPE CASSETTES

When some early programmable controllers were using PROM for program storage, others were using magnetic recording tape. This tape is very similar to the tape used in tape recorders. Since the program is a combination of high and low signals (on and off), it is easily stored on magnetic tape. The only problem this presents is that tape can also be easily altered by electrical noise and magnetic fields from large motors and transformers. This became a difficult problem because the tape recorder needed to be near the processor to record the tape. This means that the tape recorder will be in the dirtiest part of the factory right beside the motor and transformer that could alter its program. This caused many industrial P.C. users to be skeptical of the safety recording tape could provide. Once the problems were identified, P.C. manufacturers began to change the way tapes were recorded, handled, and stored. For instance, the tapes were placed in cassettes. These cassettes were similar to types used for recording music, except the industrial grade cassette has a metal case and better quality tape. The cassette recorder was made from durable plastic or metal so that it would hold up on the factory floor. These improvements along with providing more capacity to store larger programs has made tape the most commonly used storage method for P.C.'s.

Some manufacturers call their tape cassettes, *data cartridges*. Nearly all P.C. manufacturers provide some means of storing the programs on tape. This chapter will not try to outline their operation, because they all tend to be a little different.

The easiest way to learn to use the cassette or data recorder is to enter a simple two- or three-line program into the processor. By using a simple program, you can make a mistake and not feel too bad that you lost or altered your program. Once you have mastered the save and load techniques, you can use any program without fear of losing it.

Most P.C.'s have a special set of instructions that appear on the CRT to help you make the recording. Usually these instructions appear in the form of a menu, which enables you to make selections or choices to connect the recorder and transfer data. Remember, since the same recorder may be used on several different brands of P.C.'s to save a program onto tape and load a program from tape back to the processor, some changes need to be made to transfer data correctly. These changes can be made from the CRT. They are called the *data transmission format*. The format enables the user to select transmission parameters so data will be transmitted between the recorder and processor correctly. Most programmable controllers and data recorders use serial transmission to transfer data. An RS232 standard is usually used. The RS232 data transmission standard allows for several transmission parameters to be selected. These parameters include speed of transmission, called *baud rate*; number of data bits; number of stop bits; and even, odd, or no parity. This probably sounds very technical, but really, the system is easy to use. In fact all you need to do is make sure that the format selected for the processor matches that of the recorder. Sometimes, either the recorder's or processor's data transmission parameters are fixed or nonselectable. This means that you need to set the other's parameters to match the one that is fixed.

It might be easier to use these parameters if you know a little about them. First of all, the data transmission rate is called *baud rate*. Baud rate is specified as the number of data bits transmitted per second. Standard baud rates are 300, 600, 1200, 1800, 2400, 3600, 4800, 7200, and 9600.

The next parameter is number of data bits. The most common number of data bit used is 7 or 8. The number of stop bits used with the data bits is usually 1 or 2.

The next parameter is parity. Parity is a method that checks the data transmission for missing or added data bits. This means that if the data is altered or lost during the transfer process, the parity check detects it. In effect, the parity adds a high bit if needed to make the transmitted data have an even number of high bits or an odd number of high bits, depending on whether even parity or odd parity is selected. If even parity is selected, the transmission is checked to be sure each group of data has an even number of high bits. If an odd number is detected the processor and recorder know the data was altered during transmission. Most recorders and processors also provide a verification function to test data transmission once it has been completed. This verification process simply checks the tape against memory.

In using the parameters, all you need to remember is to make sure the processor and recorder are set the same. This means their baud rate, data length, stop bits, and parity must match. Most data recorders and processors have easy-to-follow user instructions making this process fairly simple.

Once you have made a tape, be sure to handle and store it carefully. Don't forget that magnetic fields from motors or magnets can alter the tape. For instance some people have reported losing tapes to some very strange sources of magnets. One loss occurred by storing the tapes in a lower file drawer, and when a large commercial vacuum cleaner was used to clean the office, a tape was altered. Generally tapes are very safe and reliable. Just be aware that they can be changed. Some P.C. manufacturers are presently using other computers, such as their IBM or DEC large system computer, to store programs. This method will someday be widely used, especially by companies that use a large number of P.C.'s.

OFFLINE PROGRAMMING

Offline programming enables the P.C.users to utilize only the industrial terminal, or keyboard, and data cassette recorder to make P.C. programs in an office, and store them on tape. When the new program is needed at the machine, the tape is used to down-load the program to the processor. After checking the program for correct operation, the P.C. can again control the machine in its new operation.

This practice enables the programmer to write new programs without taking the processor out of commission. Offline programming was not possible with the earliest P.C. systems. This meant that the program could only be written into a processor. Since the processor was busy executing the present machine program, it meant that the machine had to be shut down to write a new program. Since some programs required hundreds of hours to write, this resulted in many hours of machine downtime. To get past this problem, industries usually purchased an additional process just to use for writing new programs. This was usually viewed as an unnecessary cost, but a necessary evil for P.C. users.

Now newer programming panels have accommodations for offline programming, which enables you to write programs even when they are not connected to a processor. Once the program is written, it can easily be stored on tape and loaded directly into the processor with a minimum amount of machine downtime.

Tapeloaders are now an integral part of offline programming and program storage. These functions make the P.C. system more reliable and cost effective.

PRINTED (HARD) COPY

Another way to document a P.C. program is to connect a printer to the industrial terminal and print a copy of the processor program. This copy is sometimes called a *hard copy*, and the copy on data cassette is called a *soft copy*. Figure 17-1 shows a picture of a typical printer.

The printed copy shown in Fig. 17-2 will look just like the electrical diagram most electricians are used to seeing. In fact most P.C. users keep a printed copy of the program right in the machine's electrical cabinet. When the machine becomes inoperative, the technician can use the printed copy to troubleshoot the system. It should be noted at this time that the printed copy provides all essential data to understand the program. If the program contains counters, timers, or math function and use registers, the printed copy identifies the registers, but may not have an up-to-date count, time, or other accumulated values stored in registers. As long as you

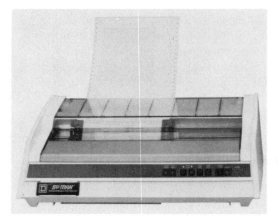

Figure 17-1 Typical printer used with programmable controllers. Picture courtesy of Square D Company.

LADDER DIAGRAM DUMP

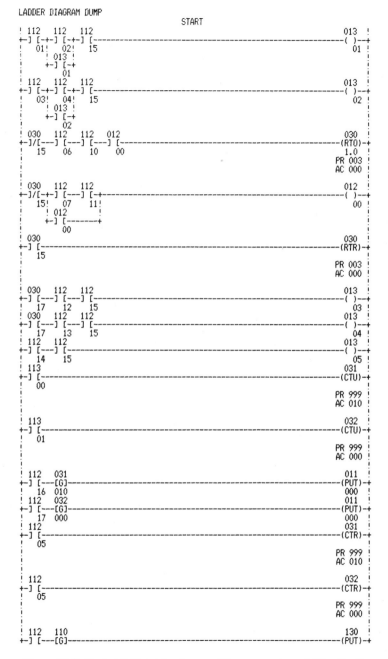

Figure 17-2 Typical P.C. ladder diagram. Diagram courtesy of Allen-Bradley Company.

understand this, you can check for current data by using the CRT and use the printed copy for all other troubleshooting.

Some manufacturers do not provide a means of printing a copy of the program in their processor. In this case the program must be transferred to some computer that can drive a printer. Usually the processors that do not support a printer have a small memory size so the program can be written down by hand.

On larger P.C. systems, it is very important that all aspects of the program are documented. This means that all counters, timers, and data functions should be accounted for, their file locations should be identified. All input modules and output modules locations should also be identified as well as the switch and electric loads connected to them. This type of documentation enables technicians to quickly locate the program rungs, machine switches and loads, and P.C. I/Os for troubleshooting and testing purposes. If program documentation does not indicate these vital locations it might take a technician hours to test larger systems.

This chapter has provided you with information concerning program storage and retrieval. It should help you understand that the P.C. system is very dependent on data stored as backup or documentation. If care is taken to store programs and document them correctly, the programmable controller system becomes a powerful control system. If programs are not stored correctly or if no printed copies are documented, the P.C. system becomes a large, continuous nightmare.

QUESTIONS

1. What is RAM memory?
2. What is meant by the statement, RAM memory is volatile?
3. What is a PROM?
4. What is the major difference between RAM memory and a PROM?
5. Explain how a PROM is erased.
6. Explain the difference between a PROM and EAROM or EPROM.
7. Name several advantages EAROM has over a PROM.
8. Name one advantage of using tape cassettes.
9. Explain the following data transmission terms: baud rate, parity, data bits, stop bits
10. Explain what is meant by the statement, The data format for the processor and storage device must match exactly for data to be transmitted.
11. What is RS232C?
12. Explain the advantage of offline programming.
13. Explain why you might need a printed copy of your program.
14. Explain the term *program documentation*.
15. Complete each of the following exercises if you have the necessary equipment available:
    (a) Make a copy of a program on a PROM.
    (b) Make a copy of a program using an EAROM or EPROM.
    (c) Make a copy of a program on a tape cassette.
    (d) Make a hard copy of a program on a printer.

chapter 18
Data Acquisition

In the past few years, more and more industries have become semiautomated or totally automated. As new products are designed, process control systems have been updated. Each of these areas of industry require a great deal of data to be analyzed for the system to operate correctly. These data include information about the system, such as temperature, pressures, flows, levels, weights, counts, times, sequences, current, voltage, phase shift, and horsepower. Many other important characteristics are also recorded so that the system can operate correctly. The job of data gathering has grown very large and complex over the last few years. The programmable controller is the ideal system to gather and act upon these data.

In order to gather data accurately the programmable controller must be able to react to data as simple as a single input from an over temperature switch, or data as complex as analog temperature recordings every 1/10 of a second. In each case, data may need to be converted from an analog to a digital signal for the processor to receive data or to act upon it by controlling an output.

DIGITAL AND ANALOG SIGNALS

A digital signal is the simplest type of signal to understand. The best example of a simple digital signal is the output of a switch. The digital signal has two states or conditions, on and off, just like the switch. In this case, only the on or off condition is recognized. Any values between on and off, such as half on, are not included in digital signals. This is why the numbering system of *1 equals on* and *0 equals off* has been adopted.

An analog signal includes all values of the signal between on and off. For instance, 1/2 or 1/3 on is a valid analog signal. A good example of an analog signal in use is the fuel gauge in your automobile. The fuel gauge registers many values between completely full and completely empty.

Both the digital signal and analog signal provide important data. The actual signal used depends on several factors. The digital signals can be grouped several ways to include placement value for digits so that a conversion between analog and digital signals can be made. These numbering systems are known as *binary* and *binary-coded decimal* (BCD). The binary and BCD systems are used to convert values between analog and digital.

DIGITAL TO ANALOG CONVERSION

Between digital and analog signals the easiest conversion to understand is the conversion from digital to analog. This is sometimes called the *D/A conversion*. The analog output signal varies from application to application. The most widely used analog current signal is 4–20 milliamps, and the most widely used analog voltage signals are 0–5 volts and 0–10 volts. Each of these analog signals varies from a minimum signal value to a maximum signal value. For the voltage signal, 0 is typically the lowest value. For the current range, 4 milliamps is the lowest value, or completely off, while 20 milliamps is completely on. The number of incremental values between minimum value and maximum value is known as *resolution*. The digital-to-analog converter takes a digital input signal and changes it to an analog output signal. The number of incremental values, or resolution, in the analog output signal is dependent upon the digital input signal. If the input signal is 4-bit binary, the output analog has 16 possible increments, including the minimum and maximum values. If the digital input is 8-bit, it has 256 increments, if it is 12-bit it has 4096 increments, and if the input is 16-bit binary, it has 65,536 increments in the analog signal. The amount of resolution over 4–20 milliamps or 0–10 volts can be very accurate if the input digital signal is 16-bit binary.

If the digital input signal is BCD, or binary-coded decimal, the 8-bit input provides 100 intervals (0–99), and if it is 16-bit (BCD) it provides 10,000 intervals (0–9999). The BCD input values are easier for operators or programmers to read, but it gives less resolution. The straight binary input gives better resolution, but it is more difficult for an operator or programmer to read. The processor does not show a preference between binary and BCD values. It uses both systems easily and accurately.

You do not need to know how the digital analog converter operates. You simply need to know what degree of resolution is needed for your application, and select the number of BCD or binary digits that will provide that resolution. The programmable controller can send a 4-, 8-, 12-, or 16-bit binary or BCD signal through its outputs to a digital-to-analog converter, or the P.C. may have an analog module mounted in its rack that converts the signal inside the module to send out a true analog signal. Regardless of whether a separate D/A converter or analog output card is used, the operation is the same.

ANALOG TO DIGITAL SIGNAL CONVERSION

The analog-to-digital signal conversion is the companion function to the digital-to-analog conversion. The analog-to-digital signal conversion, sometimes called *A/D conversion*, is used to convert an analog input signal to a digital output signal. The typical analog input signal is 4–20 milliamps or 0–10 volts. This is similar to the analog output signal values. The 4–20 milliamps or 0–10 volt input signals can have an infinite number of increments between the minimum and maximum values. The number of binary or BCD digits in the output actually determines how many of the increments are useful. For instance, if the output is 8-bit binary, 256 increments are used, and if 16-bit binary is used 35,536 are used. If 8-bit BCD is used, 100 intervals (0–99) are usuable. You do not need to understand the actual operation of the A/D converter, instead all you need to understand is that the input analog signal varies from a minimum value to a maximum value and the number of increments depends on the number of binary or BCD digits used in the output.

The A/D converter is generally used to convert an analog signal to digital for a part of the system that can use only digital signals. For example, some small P.C. systems cannot support analog modules directly, so the signal is sent to an A/D converter, and the converted digital signal is then sent to the P.C. input module.

Both A/D and D/A converters have become more reliable and inexpensive. Today they are used as part of modules or independently within the P.C. system with a high degree of reliability and accuracy.

WORD TRANSFER (BLOCK TRANSFER)

The Allen-Bradley P.C. system provides several functions that enable the processor to move data around in memory for storage and other functions.

These instructions enable the movement of data. The block transfer function can be used to input binary or BCD data from an analog thermocouple or other modules. This data can be stored in a file for report generation or compared to data in other files.

The BCD input requires 12 bits for each three digits. This means that a 3-digit BCD input needs one 16-bit data word. The files need to be large enough to store the 16-bit data words. Remember the term *word* is used to describe a group of 16 bits of data that can represent a temperature, pressure, or some other analog value.

Allen-Bradley uses other instructions, besides the block transfer functions, to transfer data. These instructions are called *GET* and *PUT*. Don't confuse Allen-Bradley's GET instruction with Modicon's GET function. Figure 18-1 shows the

Figure 18-1 Allen-Bradley GET and PUT functions. Diagram courtesy of Allen-Bradley Company.

GET and PUT functions used in Allen-Bradley's program. The GET instruction causes the processor to go to the register indicated in the GET instruction and move the 16-bit data word to the register indicated in the PUT instruction address.

The GET instruction is logic dependent. Figure 18-2 shows a typical GET and PUT circuit. When the enable switch is energized, the GET instruction and PUT instruction are executed. If the data in the data table are BCD, 3 digits or 12 bits are stored in the 16-bit word. The last four bits are not used. Any time a new data word is received at a register it writes over the previous value in the register. The block transfer function and the GET and PUT instructions enable the programmer to construct very complex program logic by moving data into and out of registers for mathematical functions and process control. The Allen-Bradley system also supports three additional file moves. These are called word to file, file to file, and file to word functions. The word to file function allows data to be moved from any word address into a file. The file to file function allows data to be moved from one file to another. The file to word function allows data to be moved out of a file to any word address.

Texas Instruments P.C. systems also support several 8-bit word and 16-bit word transfers. These instructions also allow the program to execute AND, OR, and exclusive OR instructions with data words. The Texas Instruments instruction set also enables the data words to be used in addition, subtraction, comparison, multiplication, division, and square root functions. These instructions enable the programmer to build complex control systems by bringing analog signals into the processor as data words. Since 4-digit BCD uses 16 bits, its input can also be brought into the P.C. as a data word. These data words can also be used as a preset value for timers and counters or for setpoints in automated control systems.

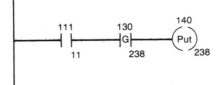

Figure 18-2 Allen-Bradley GET and PUT functions in a program. Diagram courtesy of Allen-Bradley Company.

MODICON BCD AND BINARY INPUT AND OUTPUT

Modicon provides several instructions that allow the processor to bring in data in BCD and binary values and transfer this data between tables and registers for storage or manipulation.

The CONVERT or CONV instruction is used to bring BCD and binary values

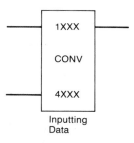

Inputting
Data

Figure 18-3 Convert used to input values. Diagram courtesy of Gould Modicon Programmable Controller Division.

Note: If lower input power node is used with upper input power node, conversions will be Binary. If only the upper node is used, the conversion will be in BCD.

into the processor's memory. These values can also be used to send BCD and binary values out to loads such as 4-digit, seven-segment LED displays. Figures 18-3 and 18-4 show the CONV instruction used as input and output BCD values.

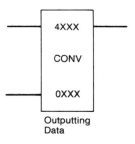

Outputting
Data

Figure 18-4 Convert used to output values. Diagram courtesy of Gould Modicon Programmable Controller Division.

Note: If lower input power node is used with upper input power node, conversions will be Binary. If only the upper node is used, the conversion will be in BCD.

Figure 18-5 shows a typical input 1020 that is converted to BCD and stored in register 4025. When the enable switch 1002 is closed the CONV function block checks the 12 consecutive inputs, starting at the input listed on the top line of the function block. These 12 digits are considered a 3-digit BCD value. In this example, inputs 1020 through 1032 are converted to 3-digit BCD. The 3 digits are converted to values 000–999 and stored in the register indicated on the bottom line of the function block. In this example, the 3-digit BCD value is stored in register 4025.

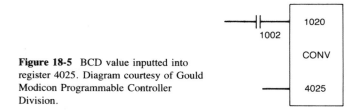

Figure 18-5 BCD value inputted into register 4025. Diagram courtesy of Gould Modicon Programmable Controller Division.

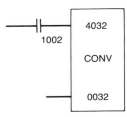

Figure 18-6 BCD value outputted from register 4032. Diagram courtesy of Gould Modicon Programmable Controller Division.

Figure 18-6 shows a typical BCD output program. When the enable contacts 1002 close, the processor goes to the register indicated on the top line of the function block and outputs its value in BCD to the 12 consecutive outputs, starting with the output listed at the bottom of the function block. In this example, the value in register 4032 is converted to 3-digit BCD and sent to outputs 0032 through 0044.

The output is usually connected to a 3-digit, seven segment display or other load using BCD as an input. Remember that the processor looks for 12 consecutive inputs and 12 consecutive outputs when the CONV function block is used. If you have 2-digit BCD input or output, the 4 inputs or outputs not used are still checked by the processor during the CONV instruction so they are unusable as regular I/Os.

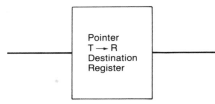

Figure 18-7 Table to register function block. Diagram courtesy of Gould Modicon Programmable Controller Division.

TABLES AND REGISTERS

Modicon also allows data to be transferred between tables and registers. Modicon identifies a table as a group of consecutive registers. In the 484 system, these can register 4001 through 4254 and any of the 30XX input registers. The data can be transferred, or moved, from a table to a register or from a register to a table. Modicon uses a function block for table-to-register and register-to-table data moves (see Fig. 18-7). These instructions also use a pointer to aid in the data transfer. The pointer is a value 1–254 or 801–832. This is pointing, or indicating, that a register 4001 through 4254 or 3001 through 3032 should be examined and that the contents, or value stored in it, should be moved. Figure 18-8 shows an example of a table to register data function block with acceptable values for the pointer and register. When the function block is enabled at the top left side, the table-to-register move is executed. The processor goes to the register indicated by the number in the top line of the function block. The content, or value stored, in this register is a number 1–254, which indicates another register 4001–4254. If a number 801–832 is stored in the register, it indicates register 3001–3032. These numbers are called the *pointer* because they point or direct the processor to yet another register.

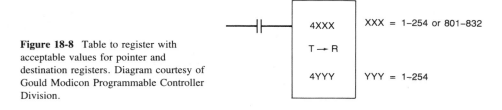

Figure 18-8 Table to register with acceptable values for pointer and destination registers. Diagram courtesy of Gould Modicon Programmable Controller Division.

The processor takes the value in the pointer register and sends it to the register indicated by the number at the bottom of the function block. The data is moved, or transferred, to the new register location. It is not changed in any way. At first this sounds very complex, but it is a very usuable technique in transferring, or moving, data. If you recall, the processor can perform mathematical functions on constant values or values found in registers. The table-to-register move enables the processor to go to tables (a group of registers), retrieve the data, and move it to a new register. This is a way for the processor to store complex data in one or more registers and retrieve it for later use, such as a comparison function. It also might help you to remember that the data being stored can be motor data, such as a full load current, voltages, and torque, or it can be the weight of material in a bin to be used in a recipe for a batch process.

The table-to-register function block provides outputs on its right side, just like other function blocks. The top output is energized if the processor finds a valid pointer number 1–254 or 801–832 in the register. If the pointer number is not a valid pointer, then the bottom output is energized. These outputs can be used to verify the operation. It provides the user with the ability to program the P.C. for complex processes and automation. Since this data is moved each time the processor scans the program, it provides constant data update.

REGISTER-TO-TABLE MOVE

The companion function of the table-to-register move is a register-to-table move. Figure 18-9 shows a typical register-to-table move. Some programmers feel that the register-to-table move must take place prior to the table-to-register move because

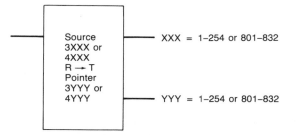

Figure 18-9 Typical register to table move.

the processor transfers the latest data it finds in the register as the scan occurs. This means that if a value in a register is storing the weight of a bin, the weight may vary as the bin empties.

The function block in Fig. 18-9 shows the format for the register-to-table move. The top line of the function block shows the register that gives the processor data. This number can be any 3001–3032 input register or any 4001–4254 storage register. The bottom number is the pointer. This is a register that has a number in its contents. The number is 1–254 or 801–832 if it is a valid pointer. These numbers represent the register that receives the data from the processor.

Figure 18-10 shows an example of this function. It may be easier to follow this process if you actually enter this program and place a value of 50 in register 4005 and a value of 80 in register 4062. Be sure to use the GET function to show the contents of register 4005, 4062, and 4080. When contacts 1001 are energized, the register-to-table move is initialized. Each time the processor scans the program, the value (50) in register 4005 is moved.

In this example the value (50) in register 4005 is shown at the bottom of Fig. 18-9 as you would see it when you use the GET function. The processor then looks into register 4062 and finds a value of 80. The value 80 is the pointer to register 4080. The processor then moves the value 50 from register 4005 into register 4080. You should see this occur in the GET area of the CRT. If you change the value or data in register 4005, the new value shows up in register 4080. Try this by moving the cursor on the CRT over register 4005 in the GET area of the screen. Now use the keyboard to put 60 into register 4005 by using the following keys: 6, 0, ENTER. When 1001 is energized, the value 60 should appear in register 4080.

If you change the value of the pointer, the value in register 4005 is moved to the new register indicated by the pointer. Try this by placing the value 85 in register 4062. Move the cursor over register 4062, and enter value 85 into register 4062. Now use the GET function to show the contents of register 4085 in the GET area of the CRT. When the function block is energized, the value in register 4005 is moved to the new register, 4085.

By now you should understand that the program needs some additional sequence or math function to cause the table-to-register or register-to-table function to operate with a series of registers or tables. With this basic knowledge you can

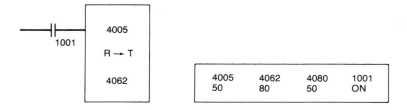

Figure 18-10 Example register to table move program with register values shown in reference area of CRT. Diagram courtesy of Gould Modicon Programmable Controller Division.

easily learn to construct such a program. You might also wonder what all this data will be used for. The next part of this chapter gives you several examples of data usage. You might also see the need for advanced intelligent modules, such as analog, TTL, thermocouples, ASCII, and PID, to send and receive more complex data. The next chapter explains the operation and function of these advanced input and output modules.

DATA BUS AND DATA NETWORKS

As programmable controllers become more sophisticated they are being used to monitor total factory automation or are being used as an integral part of the automated system reporting to other computers. The exchange of data between P.C.'s on the factory floor and master P.C.'s in a control room or other host computer make up a data network. Most major P.C. manufacturers provide some type of network capability for their systems. Some manufacturers even design the data transmission

Figure 18-11 Typical communications module. Picture courtesy of Allen-Bradley Company.

format and give their system a name. For example, Texas Instruments' data network is called the TIWAY; Modicon calls its network, MODBUS; Allen-Bradley calls theirs the Data Highway; and General Electric calls their system the GEnet. These data networks can become very complex. For this reason, the following material is of an informative nature rather than technical matter. The data networks generally allow P.C.-to-P.C. communication, which is considered low-level communication. Usually one P.C.in the system is configured as a master, and the other P.C.'s are called the *slave system.* Sometimes this is called peer-to-peer communication or expansion of the normal P.C. system. The total number of P.C. units connected together is generally limited but extensive enough to allow for total system monitoring. Figure 18-11 shows a typical module used to connect P.C.'s in this type of data network.

TRANSFERRING DATA FROM P.C. TO MASTER AND MAINFRAME

Once data is put on the data network it can be sent to a host P.C. somewhere inside the factory or to a mainframe computer in another state. Usually traditional high-speed modems and regular telephone lines are used for data transmission. Figure 18-12 shows a block diagram of Texas Instruments' TIWAY network system. You should notice that P.C.'s like their 5TI, TI510,T520, TI530, and PM550 can all send and receive data from the network. This data may be sent to a control center inside the factory, which may include a printer for report generation and a color graphics system for a complete update.

Figure 18-13 shows a typical MODBUS data network system, including MODWAY used by Modicon. The MODBUS network allows the network to accept and send data to any Modicon 884, 584, or 484 P.C. in the system. This data can

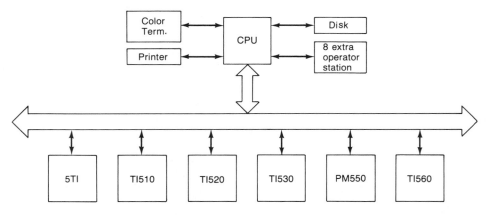

Figure 18-12 Typical layout of TIWAY data network. Diagram courtesy of Texas Instruments.

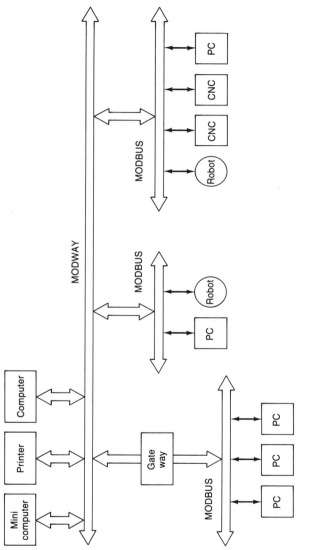

Figure 18-13 MODBUS and MODWAY data network system. Diagram courtesy of Gould Modicon Programmable Controller Division.

be transmitted to any mainframe computer located onsite or remotely at headquarters. The MODWAY system also allows data to be transmitted between P.C.'s that are not manufactured by the same company.

Figure 18-14 shows typical modules used to transmit and receive data from the MODBUS system. This means that if a company has P.C.'s with several different brand names, they can all transmit data on the network. This also allows for communications between P.C.'s and robots or other high-technology devices, such as CAD-CAM systems.

You should understand that this is the newest area of technology for P.C.'s. It is also such a large, complex volume of information that it would take several textbooks to completely cover the subject. For the most part, every major P.C. manufacturer has some means of transmitting data between systems. As you reach the level of using and needing network information, you should consult with the field technician or engineer for your P.C.

Figure 18-14 Typical modules used for the MODBUS system. Diagram courtesy of Gould Modicon Programmable Controller Division.

TYPES OF DATA TRANSMITTED

By now you are probably wondering about the data that is being transferred from P.C. to P.C. and to host computers at headquarters. Today assembly industries are switching to a limited inventory supply. This system is called *just-in-time delivery*. For instance, automobile assembly factories keep just enough parts in inventory to make automobiles for one day or three to four shifts. This saves a huge amount of money in warehousing and parts handling. As parts are needed, they are shipped directly to the point on the assembly line where they are installed. This system is very efficient, as long as just the right number of parts are kept on hand. If too few parts are shipped or if too many arrive, the system becomes a nightmare.

A master computer is used to schedule parts shipments. Since a variety of parts are needed from several sources, the computer can send a production order directly to the P.C. that is controlling the machine making the auto parts. As the machine makes parts during each shift, the P.C. reports back to the master computer exactly how many parts have been completed. As a deadline approaches, the master computer can decide if the parts are being produced fast enough to meet shipping schedules. If a problem develops, such as a broken machine, downtime or shipping problems, the master computer can make adjustments by calling on another machine, possibly at another site, to help fill the orders. This application is rather typical of data sent on networks. Usually network data contains production reports, quality control reports, energy usage, and shipping reports that concern headquarters and remote sites.

In essence, the data network enables the P.C. system to become a full partner in automation. The P.C. system can monitor itself and send reports about maintenance and operation for in-plant usage and more complex reports for corporate use. It is also noted that you may need to study some data transmission systems, such as RS232 serial or IEEE parallel interfaces for additional information. Questions at the end of this chapter will help you fully understand this material.

QUESTIONS

1. Explain the difference between an analog signal and a digital signal.
2. Give an example of an analog signal.
3. What is a D/A converter?
4. How many analog values can you possibly get from an 8-bit binary signal? From a 12-bit binary signal?
5. How many binary-coded decimal (BCD) values can you get from 8 bits? 12 bits?
6. What is an A/D converter?
7. Explain how the Allen-Bradley GET and PUT instructions operate.
8. Explain how the Modicon Conv instruction operates to input a data word and output a data word.
9. Explain what a pointer is and how Modicon uses it in its table-to-register moves.
10. What is a register?
11. What does Modicon call a table?
12. Explain what a data network is.
13. List the names of the data network for G.E., Modicon, Allen-Bradley, and Texas Instruments programmable controllers.
14. Explain how P.C.'s made by different manufacturers can be connected to the same data network.

chapter 19
Advanced I/O Modules

All programmable controller manufacturers have been working hard to provide a variety of modules. These newest modules include analog input and output modules, thermocouple modules, TTL modules, real-time clock, ASCII modules, encoder/counter modules, and PID modules. Each of these modules was designed to meet certain needs of larger, more complex P.C. systems. You may remember that the first P.C.'s were built to replace relays. These systems used simple input and output modules that operated from various voltages of AC and DC electricity. As these P.C. systems became more complex and new applications were found, the need for more varied types of I/Os was discovered. From these needs, new, reliable, and complex modules were designed. Another breakthrough evolved when microprocessor chips became smaller, more reliable, and less expensive and were capable of operating inside modules. Modules that have microprocessor chips on-board have been termed *intelligent modules* because the microprocessor chip controls the module and reports to the P.C. Not all of these new modules have microprocessors on-board. Some of the modules are using the latest operational amplifier, or op-amp, technology rather than microprocessors to carry out complex control operations. This chapter discusses the basic operation of some of these advanced modules.

ANALOG MODULES

One of the more usuable advanced modules is the analog module. Both input and output analog modules are available. Traditional controls have used analog values to send and receive minimum through maximum control signals. Digital signals, on the other hand could only vary the extremes, off and on. For example, in a digital system, a bin level or tank level is indicated as empty or full.

If an analog level control is used, the exact amount of material in the bin can be determined. If a digital (on/off) signal is used to control temperature, the only values that can be indicated are too hot or too cool. If an analog signal is used it can determine each temperature along the range.

Typical output analog signals are used to drive motorized values open and closed, and to operate all types of electrical drives, such as SCR, TRIAC, and transistors. These electrical drive systems are used to control electric furnaces, AC and DC motors, electric valves, and servomotors. If a digital signal is used to control a motor, it can only be on or off. If an analog signal is used the motor can be controlled from slow speeds up to full speed. Analog-controlled furnaces can control input heat to vary temperature over a wide range, from cool to very hot.

A typical analog input module is shown in Fig. 19-1. Even though the picture is of an Allen-Bradley module, most all major P.C. manufacturers have similar modules. The analog input module is configured with dip switches inside its cover. These dip switches enable the user to set up the analog module as a stand-alone module or as a master module to be used with an expander module. The stand-

Figure 19-1 Allen-Bradley analog module. Picture courtesy of Allen-Bradley Company.

alone module can control eight separate analog channels. If an expander is added, it allows more channels to be multiplexed with the same analog input module. By multiplexing, the processor can accept more analog data through one module.

Other dip switches set the module for voltage or current type analog signals. This gives the analog module the ability to fit with existing analog signals. If these signals use voltage, the module can be set to receive analog voltage, and if the signals are presently current, the module can be set for current signals. This also enables the P.C. manufacturer to make one analog module to fit a variety of situations that are found in industry.

Typical DC voltage options are 1–5 volts, 0–10 volts, 0–5 volts, and $+/-10$ volts. The typical current options are 4–20 ma, 0–20 ma, and $+/-20$ ma. These options enable the module to be configured to accept any analog signal that may be found in industrial or process controls.

Another set of dip switches enables the signal to be sent in BCD or straight binary when it is transmitted from the input analog module to the processor. All of these options provide a means for the module to be fitted, or adjusted, to the system. This also means that one analog input module fits all applications.

ANALOG OUTPUT MODULES

The analog output module is very similar to the analog input module except it sends a true analog voltage or current signal out to the control devices, such as a motor drive, temperature control, or motorized valve. The processor in the P.C. sends a digital binary or BCD signal to the output module. The output module then converts the digital signal into an analog output signal. This analog signal is then sent to the control device connected to the output analog module's terminal. Most analog output modules can drive multiple channels. These modules can also control six to eight channels or separate analog output devices. If an expander module is used, more channels can be controlled by allowing the processor to multiplex the signal to the expander and analog modules. It is also important to understand that several analog output modules can be controlled together as a system of master and slaves.

The analog output signal is sent from the processor to the analog module through a block function or a data file move. Usually the output for one channel will be controlled by 12-bit data words or 3-digit BCD. This means that the data word can be broken down to correspond to the 4–20 ma output signal. Again you should remember that the number of incremented values in the analog output signal is 4096 for a 12-bit binary signal and 999 for a BCD controlled signal.

The analog output signal can be current or voltage. The voltage signal can be 0–5 volts, 1–5 volts, 0–10 volts, $+5$ to -5 volts, and $+10$ to -10 volts. The current signals can be 4–20 ma, 0–20 ma, and -20 to $+20$ ma.

Another feature of the analog output module is that the gain of the output signal can be selected through dip switch settings. Some other modules may use jumpers, or wire wrap, to configure the module to fit the system.

Analog input and output modules have provided P.C.'s a direct link between digital and analog systems. Most manufacturers' modules are capable of being configured during installation to match the system. This allows the programmer and installation technician a great deal of freedom in making systems compatible. Analog modules have become the backbone of systems that control processes in manufacturing. These modules also make the P.C. capable of receiving vital process data and make decisions for outputs. The analog modules are generally used in conjunction with other intelligent modules and data processing modules.

THERMOCOUPLE MODULES

Another data gathering module available from most P.C. manufacturers is a thermocouple module. Figure 19.2 shows a picture of a typical thermocouple module. This module is manufactured for the Allen-Bradley P.C. system. Other manufacturers provide similar thermocouple modules. This thermocouple module provides eight channels that allow the module to take data from eight different thermocouples. The thermocouple module can also be coupled with an expander module that allows

Figure 19-2 Allen-Bradley thermocouple module. Picture courtesy of Allen-Bradley Company.

the P.C. to multiplex more thermocouples through the thermocouple module and expander. This allows the system to check a large number of thermocouples at a relatively inexpensive price.

Thermocouple modules typically have dip switches or jumpers to configure the module during installations. The dip switches allow the thermocouple module to stand alone or in a master, slave system. Other switches allow the user to set the temperature measurement in degrees Fahrenheit or centigrade and the data format to binary or BCD. Some modules also allow some type of scaling of larger temperatures, such as divide by 2, inside the module, while other systems do all scaling inside the processor.

Switches may also be provided to select the type of thermocouple being used, such as type J, type K, type T, type E, type B, type R, and type S. Another switch or jumper may provide high or low temperature ranges. Each of these selections provides the P.C. system with a wide variety of options to fit most any thermocouple system presently installed or being installed. The thermocouple module allows the processor to check temperatures thousands of times per second. Each of the temperatures can be stored as data or used for comparison or other data manipulations. Some earlier P.C. systems needed to use A/D converters and analog modules as descrete systems to provide thermocouple inputs. This was found to be expensive, slow, and too complex for today's control systems.

Modern thermocouple modules provide fast and accurate means to gather temperature data. The P.C. program can be written to input data words during each scan, and this data can be used for total control.

PID MODULES

The most complex module available for P.C. systems is called the PID module. PID stands for Proportional, Integral, and Derivative. These terms represent process control and instrumentation operations. Figure 19-3 shows a picture of a typical PID module with other intelligent modules. PID is a method of controlling complete loop systems, such as temperature, pressure flow, and fluid levels. The PID operations are complex and mathematically based. Because of the complexity, an engineer usually designs the PID program control. This program is usually called an algorithm. An algorithm is a complex program based on mathematical calculations.

The PID module allows the process control to take place at the module rather than burden the processor with continual control updates. Basically, the PID module has a microprocessor chip that works similarly to the processor on-board the P.C. The processor in the PID module is set up to take in analog data signals from various transducers, such as temperature, pressure, or flow. The module compares this data to set points provided by the processor's program and determines the appropriate output signal. Since the PID module only needs to update the processor periodically, the PID module can provide continuous control of the input and output signals without slowing down the processor scan.

Figure 19-3 PID and other intelligent modules. Picture courtesy of Allen-Bradley Company.

The PID module is able to handle several channels or control loops, and the module is totally configured at the time of installation. The configuration includes types of analog signals, types of digital signals, and other signal conditions. The module can also be configured to only the proportional operation, or the proportional and integral operation, as well as total PID. This allows the programmer and system designer to use only the part of the PID control, as required by their application.

The processor inside the PID module may not be able to perform all complex algorithms in pure mathematical form. For this reason, some processes can not use the PID module for control. For these control systems, the PID must be performed by the computer. The Texas Instruments PM550 and TI 560/565 are designed to do these complex PID calculations in the processor in pure mathematical form. This process control is more accurate but also requires more processor time, which slows the program scan time slightly.

The PID control provided by the module tends to be less expensive than computer-operated systems. The decision concerning which system will be usable for your application must be determined by process engineers and control engineers. Both systems provide the P.C. overwhelming control capabilities for process control functions.

The most important feature about the PID module is its use of a separate on-board microprocessor to take some of the work off the main P.C. processor. This is the current trend for P.C. manufacturers as P.C. systems become larger and more complex. Other intelligent modules are being developed to provide other controls for the P.C.

OTHER INTELLIGENT MODULES

Intelligent modules are modules that contain on-board processors or other high-tech circuitry. Some examples of these modules not discussed yet include ASCII modules, which provide data transmission between the processor and printer for report generation or between the processor and data message displays; stepper modules

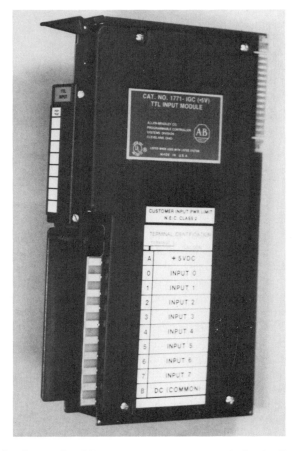

Figure 19-4 Allen-Bradley T T L module. Picture courtesy of Allen-Bradley Company.

Figure 19-5 Texas Instruments intelligent specialty modules for ASCII reports. Picture courtesy of Texas Instruments.

or positioning modules for robotic applications; and a real-time clocks module, which provides accurate timing that does not depend upon processor scan times. The TTL module is another advanced module. This module conditions the signal to TTL levels.

Each of these modules is designed to do more of the work right in the module, thus freeing the processor for program scan and I/O execution. A large variety of these modules is available from each P.C. manufacturer. This means that most P.C. brands are able to support your particular applications. In some cases, one major brand provides a particular intelligent module that is not available from some other company. Generally, each company is staying abreast of industrial needs and has modules available in developmental stages to fit the new application. Several of these modules are pictured in Figs. 19-4, 19-5, and 19-6.

Intelligent modules determine how much work can be diverted from the processor in complex control systems. Older P.C. systems tended to rely on the main processor to do all data gathering, storing, and manipulation, which slowed the scan time and limited the P.C. size. Newer systems are using more intelligent and advanced I/O modules that take this work from the main processor and allow it to scan the program and update it more often. As new modules are developed,

Figure 19-6 Allen-Bradley closed loop servo module. Picture courtesy of Allen-Bradley Company.

you will need one or two days of training to fully understand their use for your application. With a solid understanding of present modules, you will find it easy to learn new ones as they are needed. Questions are provided at the end of this chapter to help you review this material.

QUESTIONS

1. What is an intelligent module?
2. What does an analog input module do?
3. List three typical analog signal voltage levels.
4. List three typical analog signal current levels.
5. List two devices that might be controlled by an analog output module.

6. Name several parameters that might be controlled or set by jumpers on the thermocouple module.
7. What does PID stand for?
8. What is the advantage of doing PID control right at the module?
9. What is the disadvantage of doing PID control right at the module?
10. Name several intelligent modules.

chapter 20
Process Control

The latest use for programmable controllers is process control. Process control is a very complex control system that involves monitoring and adjusting instrumentation and controls of processes used to produce a product. Typical process control systems can be found controlling temperatures, pressures, and other conditions in refinery operations, chemical plants, and food processing. Process control is usually used where the process variables, such as temperature and pressures, must be strictly monitored and controlled.

It may be easier for you to understand process control when you have looked at an example. Figure 20-1 shows a block diagram of a closed loop system that makes up a small portion of the total process control system. The input portion of the diagram represents some demand placed on the system. Basically the process material, such as the fluid, flows directly from the input, through the control element, and to the output. The remainder of the diagram represents the movement of the control signal. The major parts of the diagram that primarily involve the signal are the summing point, control element or measurement element, controller or P.C., and feedback signal.

Basically the program in the P.C. defines the setpoint. The measurement element begins reporting the temperature to the processor. The processor compares the present measurement signal to the setpoint. Any difference between the present measurement signal at the output and the setpoint is considered a process error. The processor then determines how much error exists and if the error shows the actual measured variable is higher or lower than the setpoint. The point where this signal comparison and correction take place is called the *summing point*. The summing point may be located in the processor or in a special control module. The

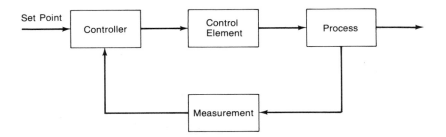

Figure 20-1 Block diagram of a closed loop system.

signal that is used to connect the output measurement signal to the summing point is called the *feedback signal*. Feedback signals provide a means to alter the process variable based on actual measurement of the output as compared to the setpoint.

These terms and this block diagram are used to explain most instrumentation and control systems. As control systems become more complex, more terms and blocks are added to this basic diagram. For instance, the amount of feedback can also be controlled by setting the gain of the signal. In fact, complex control systems can be figured mathematically or complex algorithms can be designed for the control system.

The P.C. is capable of sampling the output temperature several hundred times a second. As the fluid temperature begins to increase, the P.C. adjusts the signal to the electric heating element to add heat until the output temperature of the fluid is at the preset level. If the fluid is being recirculated, the temperature increases gradually up to the setpoint, and the amount of heat added is diminished. If the fluid is only moved through the heater once, the amount of heat needed to increase the fluid temperature stays rather constant, depending on the fluid temperature prior to reaching the heating element.

You now should have an idea of what process control is. In larger systems, the process control actually controls every parameter involved in the system. In the previous example, the other variables might be fluid flow, fluid pressure, and fluid level. The P.C. provides a control loop for each variable and controls each one independently while it checks the total system's operation.

CLOSED LOOP SYSTEMS

The system described in Fig. 20-2 is called a *closed loop system* because the fluid temperature at the output is measured and compared to a setpoint in the P.C. program. If there is any difference between the setpoint and the output temperature of the fluid, the processor makes changes in the amount of heat the heating element puts into the system. If the output temperature is warmer than the setpoint, then the P.C. turns the heating element off completely.

A basic block diagram of a closed loop system is shown in Fig. 20-2. Notice

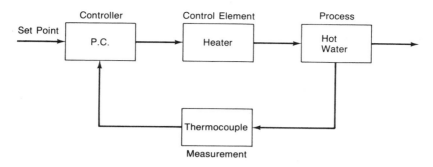

Figure 20-2 Block diagram of a typical temperature loop.

this diagram looks similar to the diagram in Fig. 20-1. This time the names of the signals will be discussed.

In the diagram in Fig. 20-2, you can see that the process variable is the temperature of the liquid that is being controlled. The controller is an electric water heating device. The measurement device is a thermocouple, and the process controller is a programmable controller. The temperature loop begins operation with the fluid flowing through the control device to the output. At a point just prior to the output, a thermocouple samples the fluid temperature and sends this data to the programmable controller, where it is compared to the setpoint. For this example let's say the present fluid temperature is 80° C, and the setpoint is 90° C. Since the fluid temperature is less than the setpoint temperature, the programmable controller sends an analog signal to the electric heater to increase the temperature of the fluid.

SETPOINT AND COMPARISON

The P.C. program must be written to periodically check the sensors for up-to-date information. Figure 20-3 shows a typical program written for an Allen-Bradley PLC-3. A picture of the PLC-3 is shown in Fig. 20-4. Other P.C. systems are programmed in a similar fashion. The first block in the program is a data input statement. In this program, the data being checked are from a thermocouple. The data word address reflects the location of the thermocouple module in the rack. The data word address also indicates the file where the temperature is being stored. These values are input and stored as BCD values. After the enable contacts are closed, the processor inputs the temperature data once every time the program is scanned. The data is stored in the data file and is also used in the comparison function block.

The comparison function block checks to see if the input temperature is less than the setpoint. The setpoint is listed as a BCD value in the comparison function. The BCD value is also stored in a data file. If a change is required in the preset value, the BCD value can be called up by the industrial terminal and changed. The

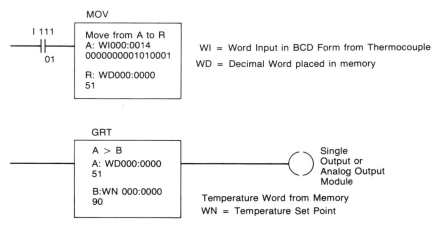

MOV

I 111
⊣├
01

Move from A to R
A: WI000:0014
0000000001010001

R: WD000:0000
51

WI = Word Input in BCD Form from Thermocouple

WD = Decimal Word placed in memory

GRT

A > B
A: WD000:0000
51

B:WN 000:0000
90

Single
Output or
Analog Output
Module

Temperature Word from Memory
WN = Temperature Set Point

Figure 20-3 Temperature comparison program.

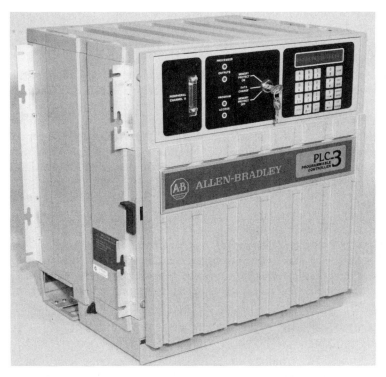

Figure 20-4 Allen-Bradley PLC-3 programmable controller. Picture courtesy of Allen-Bradley Company.

comparison can be made to see if the input temperature is less than or greater than the setpoint. In either case, the output can be controlled to add any amount of heat to bring the temperature equal to the setpoint. If the temperature is equal to the setpoint, the output can be turned off until more heat is needed. The output data word also shows the address of the analog ouput module where the heating element is connected. The output data word is also in BCD value, and the technician can check the output value to determine how much heat the heating element should be producing. Since the final control element is controlled by an analog signal, it can produce any amount of heat input between full on and full off.

COMPLEX CONTROL SYSTEMS

So far, the examples used to explain the control system have involved just one variable. In reality, most systems actually sense and control several variables. In some cases, like temperatures, the system may use dozens of thermocouples to check temperature all along the system. The P.C. still only controls one final control element for each process variable, even though the system is sensing many parameters of the system.

Another condition must also be considered when one process loop may affect another. For instance, if the P.C. is controlling a temperature, flow, and pressure loop for a fluid process, each of the control variables has an effect on the other. The temperature increases if the fluid flow is slowed through the heating element. Conversely, if the flow is increased, the temperature of the fluid decreases if the same amount of heat is added by the heating element. The relationship between the other variables is such that if the pressure is increased, the temperature increases and the flow decreases. You can see that the programmable controller must sample temperature, flows, and pressures quite often to determine the effects caused by changes in any one of the variables. Since the P.C. can sample and make comparisons thousands of times per second, it can alter each variable while checking for effects on the other variables.

The P.C. can also store data concerning the system's operation while the process variables are changed. A technician can try several programs to find the ultimate range, gain feedback error, and other variables to make the system run smoothly and accurately. Once the best program variables are determined they can become part of the permanent program. This allows the P.C. to help determine best control model for an operation. System response can also be stored and checked after several months, and minor adjustments can be made if necessary.

SETTING INNER AND OUTER LIMITS

Another type of programming for process control uses inner and outer loops for setting multiple setpoints. These setpoints can indicate when a process is reaching a point that is becoming unsafe. For example, if the process is supposed to operate at 80° C and 100° C is deemed too hot, the inner loop setpoint may be set at 90° C. If the process temperature reaches 90° C, an alarm can be used to tell the operator or technician that the process is getting too warm. If the system does not automatically start to cool down and the temperature continues to climb, the system begins to shut down when the temperature exceeds the outer limit preset temperature. For this example the outer limit could be set at 100° C. Figure 20-5 shows a typical program that can be used to program these inner and outer limits for the temperature loop. You can see from this program that these setpoints only protect against temperatures that become too high. A similar program is needed for temperatures that become too low.

The program in Fig. 20-5 is written in generic terms so you need exact addresses for I/Os and word files to make it operate. The terms in each function block help you see where preset values and actual measurements are used. The first function block shows the thermocouple temperature being input through a block function. The data word represents the actual process temperature. Remember, a data word is 16 bits of data. The temperature data word is changed to a BCD value in the second function block. The temperature data word is in BCD format for easy reading. The BCD temperature is also used in each of the comparison blocks.

In the first comparison function, the BCD temperature word is compared to a setpoint of 90. Since both values are in BCD, a direct comparison can be made. This first comparison represents the inner loop. If the process temperature exceeds 90° C, the function block energizes its output. This output is used to energize an alarm message and alarm signal. The alarm message is automatically sent to the printer, where it would appear as a printed message. The message could read: THE TEMPERATURE OF THE PROCESS HAS EXCEEDED 90°C. CHECK THE TEMPERATURE LOOP FOR MALFUNCTION. IF THIS CONDITION IS NOT CORRECTED THE PROCESS WILL BE SHUT DOWN WHEN THE TEMPER-ATURE REACHES 100° C.

The output could also be used to energize a warning buzzer or light, indicating the system is not operating correctly. When the operator or technician hears or sees the warning, they could go to the printer to see which part of the loop is malfunctioning.

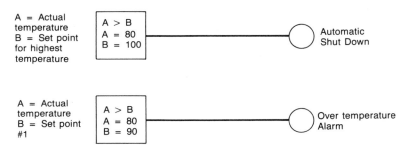

Figure 20-5 Program of inner and outer limits. Diagram courtesy of Allen-Bradley Company.

You must remember that the input temperature data word is also being used in another function block in the program for controlling the heating element. As discussed earlier in this chapter, when the temperature increases beyond the operating setpoint of 80°, the P.C. sends an analog signal to the heating element control to turn down the heat. If the temperature continued to rise, it would indicate some malfunction or fault has occurred that requires immediate attention.

If the temperature continues to rise above 90° the second safety alarm is energized when the temperature becomes higher than 100° C. The second alarm is the outer loop of the safety circuit. The output of the second comparison would be used to begin emergency shut-down procedures. These automatic shut-down operations are written into another part of the program. The safety outer loop would indicate an unsafe condition, then shut the process down if the condition did not get better.

You should begin to see the strengths of the programmable controller in this type of control system. Figures 20-6 and 20-7 show pictures of the Modicon 584L and the Texas Instruments TI560/565 processor, which are especially designed for process control.

Figure 20-6 Modicon 584 L programmable controller. Picture courtesy of Gould Modicon Programmable Controller Division.

Figure 20-7 Texas Instruments Model 560/565 programmable controller. Picture courtesy of Texas Instruments.

For instance, once the process temperature is brought into the P.C. as an input data word it can be converted to BCD and then used in as many function or data blocks as necessary. In this program, the temperature data word is used three different times. It is used for the inner loop comparison, the outer loop comparison, and the operational loop. It could also be used as a storage word where temperatures can be stored for quality control data. These temperatures could be sampled at any time interval chosen by the programmer up to the speed of the processor scan. This would give the operator a good idea of how well the system maintained the preset temperature.

Another strength of the P.C. used in this type of control is if a different product is produced in the same system on another day. The second product may be similar but have different inner and outer operating temperatures. The operator can call up a totally new program or just adjust the preset temperatures in the three comparison function blocks with the new programming panel or with a set of thumbwheel switches. The program can be written so the thumbwheel switches can address each function block and the operator can dial in a new preset value.

The only limiting factor in this control system is how complex and automated you want the system to become.

COLOR GRAPHIC SYSTEMS

Another feature being added to most P.C. systems for process control is color graphics. Figure 20-8 shows a picture of a typical color graphics system used with the P.C. The color graphics system consists of a terminal and CRT that allows the system to be programmed with typical graphic symbols and colors that represent your actual process and layout. For instance, the temperature loop may be programmed in red and the pressure loop in blue. Once the color graphics system is programmed it follows the system's operation by highlighting each section.

Figure 20-8 Color graphics system. Picture courtesy of Square D Company.

Basically the color graphics system has been designed to read input data from the P.C. processor and translate that data into system operations that are graphically displayed on the color CRT. This means an operator or technician needs a graphic T.V. picture to follow the process. The inner and outer loop setpoints and operational data can also be displayed to show present system status. When a fault occurs, the color graphics system can indicate which loop is in trouble and where the problem may be.

Color graphics is adding a visual dimension to a complex systems of pipes, controllers, and electronics that may be spread over several acres. In this way, the complete system can be safely monitored and controlled some distance from the system.

Another system available for use on some P.C. systems, in addition to color graphics, is an interactive voice system. The system essentially indicates process status with a voice synthesizer. This system gives the operator an audio indication as well as a visual one. Sometimes the system is so complex, the operator cannot monitor all status indicators or visual graphics. The audio system responds, allowing the operator to concentrate on visual indicators while hearing the audio status report in plain English. This also enables the operator to respond to situations without neglecting other process indicators.

By now you can see just how versatile the programmable controller can be. It can be set up to completely control one or two inputs and outputs, or it can be expanded to control complex process control and automation. Hopefully you now understand how the largest P.C. system is only an expansion of basic operations. Once you have mastered the basic P.C. operations, you can begin to plan and program complex P.C. systems.

QUESTIONS

1. Explain what process control is.
2. Name the four parts of a closed loop system.
3. Explain how a closed loop system differs from an open loop system.
4. Explain what a setpoint is.
5. Explain what the MOVE function does in Fig. 20-3.
6. Explain what the COMPARISON function does in Fig. 20-3.
7. What is the setpoint temperature in Fig. 20-3?
8. Explain the idea of using inner and outer limits to control a system.
9. What will happen if the inner limit of 90° is exceeded in Fig. 20-7?
10. What will happen if the outer limit of 100° is exceeded in Fig. 20-7?
11. What is meant by the term, *temperature data word*?
12. In Fig. 20-3, what is the actual temperature that the temperature data word represents?
13. Explain how a color graphics systems can help in a process control system.

Index

A

Accumulative:
  count, 71
  time, 54
  value, 54
A/D conversion, 173
Adapter module, 111
Add function:
  Allen-Bradley, 98
  Modicon, 87–88
Advanced I/O modules, 184
Algorithms, 195
All clear, 33, 35
Allen-Bradley:
  GET, 98
  history, 12
  numbering system, 44
  sequencer, 80
Analog:
  modules, 184
  output modules, 186
  signal, 172

AND Gate, 129
ASCII, 184
Assembly area, 34
Automatic resetting timer, 61

B

Back-up battery, 164
Base eight system, 24
Base two number system, 22
Baud rate, 167
BCD (Binary-coded decimal):
  input, 174
  numbering system, 23
  signal, 175
Binary numbering, 22
Bit examination, 149
Block transfer, 173
Blown fuse indicator, 148
Burning a prom, 165